T0188769

An Introduction to the Philosophy of Time

An Introduction to the Philosophy of Time

SAM BARON AND KRISTIE MILLER

polity

First published in 2019 by Polity Press

Polity Press
65 Bridge Street
Cambridge CB2 1UR, UK

Polity Press
101 Station Landing
Suite 300
Medford, MA 02155, USA

ISBN-13: 978-1-5095-2451-8
ISBN-13: 978-1-5095-2452-5(pb)

A catalogue record for this book is available from the British Library.

Library of Congress Cataloging-in-Publication Data
Names: Baron, Sam, author. | Miller, Kristie, author.
Title: An introduction to the philosophy of time / Sam Baron, Kristie Miller.
Description: Cambridge, UK ; Medford, MA : Polity Press, 2018. | Includes
    bibliographical references and index.
Identifiers: LCCN 2018011849 (print) | LCCN 2018040377 (ebook) | ISBN
    9781509524556 (Epub) | ISBN 9781509524518 (hardback) | ISBN 9781509524525
    (pbk.)
Subjects: LCSH: Time.
Classification: LCC BD638 (ebook) | LCC BD638 .B3369 2018 (print) | DDC
    115--dc23
LC record available at https://lccn.loc.gov/2018011849

Typeset in 11.25 on 13 pt Dante
by Fakenham Prepress Solutions, Fakenham, Norfolk NR21 8NL
Printed and bound in the UK by CPI Group (UK) Ltd, Croydon

The publisher has used its best endeavours to ensure that the URLs for external websites referred to in this book are correct and active at the time of going to press. However, the publisher has no responsibility for the websites and can make no guarantee that a site will remain live or that the content is or will remain appropriate.

Every effort has been made to trace all copyright holders, but if any have been inadvertently overlooked the publisher will be pleased to include any necessary credits in any subsequent reprint or edition.

For further information on Polity, visit our website: politybooks.com

Contents

Acknowledgements

Parts of this book have appeared elsewhere. Figures 19, 20 and 21 have previously appeared in S. Baron, 'Back to the Unchanging Past', *Pacific Philosophical Quarterly* 98 (2017): 129–47. Chapter 4 has appeared in an adumbrated form in S. Baron, 'Time, Physics and Philosophy: It's All Relative', *Philosophy Compass* 13 (2018). The authors are grateful to *Pacific Philosophical Quarterly* and *Philosophy Compass* for allowing the re-use of this material.

In addition, Kristie Miller would like to thank Annie and Freddie Braddon-Miller for their advice throughout this project. Sam Baron would also like to thank Sara Linton for her support and would like to note that Sara did not, in fact, kill her plants by failing to water them as Chapter 6 of this book would seem to suggest. Sara is happy to report that her plants are alive and well.

List of Figures

Introduction

Time is woven into the fabric of our lives. Everything we do, we do *in* and *across* time. It is not just that our lives are stretched out in time, from the moment of birth to the moment of our death. It is that our lives are stories. We make sense of ourselves, today, by understanding who we were yesterday, and the day before, and the day before that; by understanding what we did and why we did it. Our memories shape our present selves, since we are the product of who we were, and who we remember being. Yet who we are lasts for but a moment. It is who we *will* be that demands attention. We are forever projecting ourselves into the future. We painstakingly plan for the needs of our future selves, all the while knowing that while what we do will make those future selves who they are, still, our future selves are elusive. So it is not simply that we *happen* to live our lives in time, the way, say, one might happen to live in Australia, or Singapore. It is hard to fathom how things could be otherwise. What would it be to live a life *without* time?

Yet time itself is intangible. Though we are keenly aware of what has been – of pain and loss, regret and pride – and of what might be – of anticipation, and dread, and longing – time itself seems unaffected by anything we do. Though we can anticipate a future birthday cake, time itself cannot be tasted. Though we can dread a future dentist's appointment, time itself cannot be touched. Though we remember the crunching of leaves this past autumn, time itself cannot be heard. How can it be that something so central to each of us, to who we are, and what we care about, can be so spectral? What, in the end, is time? Is time real, like a substance in which each of us, like insects in amber, finds ourselves trapped? Is time a dimension through which we can move? Is time not a *thing* at all, but some connection between events? If time is real, what is it like? Does it flow like a river, taking each of us prisoner in its current, washing us ever closer to the end of our lives? Is time like a cresting wave, coming ever closer to each of us from the future, engulfing us, and then receding into the past? Or is time static? Like a one-way road that each of us might wander, with no chance to turn back? Or perhaps time is like none of these things. Perhaps time is more like space; something we can navigate freely, so that we can visit different times just like we visit different places. Perhaps we can travel in time the way we can travel in space.

These questions, and many more, are the focus of this book. Now, you might wonder what philosophers are doing writing a book about time. Surely that's a job for physicists! It is true: a great deal of what we know about time stems from our understanding of physics. However, even in physics we find that there are questions about time that go unanswered. Moreover, increasingly, physicists are turning to philosophers for help in understanding the role that time plays in current physical theories. Our goal is to get the reader up to speed on some of the central issues that influence our understanding of time so that they too may play a role in this ongoing discussion between science and philosophy. We hope that in doing so it will become clear what philosophy can bring to the study of time.

There is a natural narrative in philosophy regarding the study of time, and we have chosen to organise this book according to that narrative. However, the book may also be approached in a 'choose-your-own adventure' spirit. While later chapters build on material introduced in earlier chapters, you can skip around, reading the book in any order you choose. For the instructor, this means that she can design her course around this book in any number of ways, depending on the particular topic she wishes to focus on.

Chapter 1 introduces some key views in the philosophy of time and sets the stage for what is to come. It distinguishes dynamic from static theories of time, and provides a sense of why one might be tempted by one or other of these views. Then it outlines various kinds of static and dynamic theory.

Chapter 2 introduces the idea of temporal flow. This is an elusive notion, and we spend some time trying to work out what temporal flow could be, and whether the notion makes any sense at all. We consider a number of ways of modelling temporal flow, as well as various objections to those models.

Chapter 3 introduces us to a central argument in favour of there being temporal flow: an argument that appeals to how the world *seems* to us. Here, we introduce the reader to the idea of temporal phenomenology, and the ways in which appealing to that phenomenology has been used to reach metaphysical conclusions. We spend some time unpicking this argument and considering ways one might respond to it.

Chapter 4 takes us into the physics of time. It offers an accessible introduction to what physics tells us about time. We focus on the connection between physics and the static theory of time, and we evaluate the claim that contemporary physics tells us that time doesn't flow.

Chapter 5 considers the question of why it is that there are so many temporally asymmetric phenomena, when according to most physicists the laws of nature are temporally symmetric. We consider a number of proposals for explaining the asymmetry, including that time flows, or that time has a primitive direction, or that time has a direction, and its having that direction reduces to something else. We also consider the possibility that time has no direction, and it just seems to us as though time has a direction.

Chapter 6 moves away from time itself to thinking about the connection between time and causation. What is causation? Is causation necessary for time? Is time necessary for causation?

Chapter 7 investigates the ways in which objects are extended across time. This chapter outlines a number of competing views about how objects persist through time. We then ask whether each of these theories is consistent with various hypotheses about temporal ontology, and about the relationship between time, spacetime and objects.

Finally, Chapter 8 delves into the weird and wonderful idea of time travel. We ask whether backwards time travel is logically possible. We talk a lot about grandfathers, and whether time travellers will ever succeed in killing their young grandfathers in the past, and we consider whether backwards time travel presents a threat to our free will.

By the end of the book you, the reader, should have a good grasp of a range of interrelated concepts in the philosophy of time, and, hopefully, a whole lot of questions about the nature of time. We don't aim to provide all the answers here (nor could we, since many questions are still open). But we do hope to provide some of the conceptual resources necessary to frame these questions, and to have whet the reader's appetite for asking philosophical questions about time.

1

Dynamic and Static Theories of Time

1.1. Dynamic versus Static Time

Time is familiar; it shapes our day-to-day lives; it provides direction and guidance to our planning; it underpins many of our attitudes to the world. And yet, despite being a fixture of everyday life, most of us couldn't offer much by way of a *theory* of what time is. Time is at once familiar and elusive. Our goal in this chapter is to provide an introduction to the modern debate about time within philosophy. We aim to equip the reader with the resources to begin to articulate, and even defend, a theory of time. We will also gesture towards some of the reasons why one might be tempted to endorse a particular theory of time; reasons that will be developed in greater detail in later chapters.

Our investigation into the philosophy of time begins with experience. Take a moment to reflect on your own experience of the world. Observe the shapes and colours around you; consider the sounds that float by on the wind; feel your own body and its ongoing process of exchange with the environment. Pause for a moment, and now search your own experiences for a sense of *time*. Is there anything in experience over and above the smells, sights and sounds that you can discern as distinctly temporal?

No? Well, try this. Meditate on your own mortality. Think about the swift hand of death that will one day come for you, as it comes for us all. Think, in particular, about the *finitude* of your own existence. One day, you will join every other organism that has existed on this planet in the dust. Now, think about all of the things that you like to do, all of the people you love and care about, and the desires you have as yet unfulfilled. There is not enough time in the world to do everything you want to do. You will die with goals unaccomplished, with words left unsaid, with love left unloved. As you ponder this, do you feel the moments begin to trickle by? Does time steal away your moments, bringing your death closer and closer?

It is this feeling, this sense that time is *passing*, pulling you inexorably towards your end, that constitutes one of the touchstones of the philosophy of time. Later, we will return to the experience of time passing and consider the extent to which it provides evidence for this or that theory of time. For now, we want you to try, as far as possible, to feel the flow of time.

By focusing on the experience of time, we can take our first steps towards developing an account of time. According to a *dynamic* approach to time, time really is as it seems to be in experience: the nature of time is fundamentally dynamic. *Time really passes*. It seems to us as though future events come ever closer until they become present, after which they recede further and further into the past. According to the dynamic theory of time, this seeming is to be taken at face value: time really does involve a kind of flow that reflects the way we experience the world.

The dynamic approach to time sits in stark contrast to a *static* approach to time. The static approach to time denies that time is fundamentally dynamic. Thus, while it may seem that future events slowly come towards us until they are present, thereafter receding into the past, this is not *in fact* how the world is. Time, on this conception, is fixed and never moves. Rather, it is we who, *in some sense*, move through time.

We will sharpen up the core distinction between the static and dynamic approaches to time in a moment. For now, we aim only to give the reader an intuitive sense of the distinction between the two views. So here's an analogy to help you develop your understanding. Consider the difference between a river and a road. A river has a current which sweeps anything caught in it downstream. A road, by contrast, has no current and does no sweeping. Rather, it is we who must make our way along the road, using whatever means of movement we might have available.

The dynamic approach to time treats time like a river. We all live our lives in the river of time, and it is the river that throws us to oblivion. The static approach to time treats time like a road. We must all wander the road, but the road itself does not move. Time, on this picture, doesn't bring death closer to us or bring us closer to death. Rather, it is we who bring ourselves closer to death by living.

1.2. Behind the Metaphor

Everything we have said so far has been couched in metaphor. While metaphors such as these are suggestive and can help to guide philosophy, ultimately one of the things that we, as philosophers, aim to do is get behind the metaphor. Doing so involves offering a theory of time. Before offering a theory of time, however, we need to take a step back and think carefully about what we are doing. What does it mean to offer a theory of time at all?

One possibility is that we are offering an account of what time is *actually* like. Another possibility is that we are offering an account of what is essential to time: an account of what it is for some phenomenon to *be time*. In order to understand this distinction we need to introduce some modal concepts. We need to distinguish between the way things are, and the way things could have been, but aren't. We can call the totality of ways things are, *the actual world*. Then we can call any complete way things could have been, but aren't, *a merely possible world*. So, for instance, there's a possible world in which you ate cornflakes for breakfast this morning, even though, actually, you ate toast. Which is to say that it's *possible* that you ate cornflakes, even though actually you ate toast.

We can now distinguish two rather different theories of time. We will call these *constitutive theories* and *extensional theories*. A constitutive theory of something aims to tell us what it is be that something. For example, a constitutive theory of dogs aims to tell us what it is to be a dog. Similarly, a constitutive theory of circles aims to tell us what it is to be a circle. So a constitutive theory of time aims to tell us what it is to be time. Translated into the language of possibility, what this means is that the constitutive theory of time aims to tell us what time is no matter how the world turns out to be. In short, we are trying to find the features that time has in every single possibility, where by 'possibility' we mean possibility in the widest sense: there is no possible way whatsoever for time to be different. This is sometimes called 'metaphysical' possibility, and is to be contrasted with 'physical possibility' which corresponds to what is possible *given the physical laws of nature*. By contrast, an extensional theory of something aims to tell us what that something is actually like. So an extensional theory of dogs aims to tell us whether there are any actual dogs, and if so, what they are like. An extensional theory of circles aims to tell us whether there are any actual circles, and if so, what they are like.

Constitutive theories of some phenomenon often aim to specify the necessary and sufficient conditions that must be met for something to be that phenomenon. A necessary condition for X's being Y is a condition that X *must* meet, if it is to be Y. A sufficient condition for X's being Y is a condition, which, if X meets it, then X is Y (but X can be Y without meeting that condition). So, for instance, you might think it is necessary for something to be a dog, that it has fur. Clearly, though, having fur is not sufficient for being a dog, since lots of things are furry, but are not dogs.

Alternatively, you might think it is sufficient to be a dog that something has a particular sort of genetic profile, and is a member of a group with a particular evolutionary history, because if something has that genetic profile and is a member of a group with that evolutionary history, then it is a dog. Notice you might think this is sufficient, but not necessary to be a dog. Imagine a wizard waves his wand and creates something qualitatively just like a labradoodle. That thing didn't evolve; it's not a member of a group with a particular evolutionary history. But you might think it's a dog nonetheless.

It is easy to see why one might think that specifying *both* the necessary and sufficient conditions to be an X tells us *what it is* to be X. Once we've sorted out all the things that are necessary for being a dog, and all the things that are sufficient for being a dog, it seems like we know what it is to be a dog. Thought of this way then, constitutive theories of time are theories of time that aim to tell us the necessary and sufficient conditions for something to be *time*.

By contrast, an extensionalist theory of time aims to tell us what time is actually like. Consider again the case of dogs. An extensional theory of dogs might tell us a lot of things about dogs that the constitutive theory of dogs does not. It might tell us where dogs are actually found, what colours they are, how they behave, how they fit into society, what they eat, and so on. For instance, imagine that, actually, all dogs are either black or white. Then an extensional theory of dogs would tell us this, since it's an important feature of actual dogs that they only come in two colours. If you see something red out in the bush, running on four legs, you know it's not a dog. But that would be no part of a *constitutive* theory of dogs. A constitutive theory would not tell us that what it is to be a dog is to be either black or white. For if such a constitutive theory were true, then any merely possible animal very much like a dog, but red, would not be a dog at all. But we don't think it necessary that dogs are either black or white, just because actually all dogs are either black or white. So that aspect of our extensional theory of dogs won't be found in our constitutive account.

A constitutive theory of time: A theory of time that tells us what it is to be time.

An extensional theory of time: A theory of time that tells us what actual time is like.

We need to be careful when evaluating different theories of time and to be clear whether we are evaluating them as constitutive theories or as extensional theories. A good extensional theory will often make a very poor constitutive theory, since many actual features of time might be features that time in fact has, but not features that time *must* have. In what follows we will focus exclusively on *extensional* theories of time. We are interested in what time is *actually*, rather than what time *has* to be like, or what it is *to be* time in the constitutive sense. But it's worth bearing in mind that many of the arguments we encounter (though by no means all) are arguments about the constitutive nature of time. These are arguments that try to show that, for instance, time *must* be dynamic, not just that *in fact* time is dynamic. And, in fact, most dynamic theorists (those who think that time is dynamic) think that this is a constitutive feature of time. They think that what it is to be time, is to be appropriately dynamic. In general, then, it's worth bearing in mind that some of the arguments for a static theory, or for a dynamic theory, are arguments that aim to show not only that time is in fact a certain way, but that time *must* be that way, while other arguments only aim to show that time is, in fact, that way.

How do we convert the static and dynamic approaches to time discussed in the previous section into *theories* of time? In some sense, this is easy to do. We can take the core aspect of disagreement between the two approaches and use it as the basis for turning those approaches into theories. The core aspect of disagreement is the idea that time passes. The dynamic approach upholds this idea; the static approach disavows it. We can, then, state a very basic extensional dynamic theory of time and an extensional static theory of time as follows:

Dynamic theory of time: Temporal passage is a real, mind-independent feature of the universe.

Static theory of time: Temporal passage is not a real, mind-independent feature of the universe.

The above statement of the difference between the static and dynamic theories of time hides a great deal of complexity. The trouble lies with the elusive concept of temporal passage that is supposed to distinguish the two views. What exactly *is* 'temporal passage'? Providing a clear and precise statement of the concept of temporal passage is difficult.

Here's one common attempt to do so: the passage of time involves some change in the present. As time passes, events that were future become gradually closer to us, passing into the present before moving towards the past and receding into distant history. This idea can be further sharpened up via J. M. E. McTaggart's three-way distinction between the A-series, the B-series and the C-series. Each of these series is a temporal ordering of times. We will first say a bit about what that means, before outlining each series and connecting it back to the concept of temporal passage.

What is a time? We will consider this issue in greater detail later on. For now, we can give an intuitive gloss on a time as follows. A time is just an instant in this sense: a full specification of everything in space at a moment. It is, if you like, a *snapshot* of the entire universe in space, with every object frozen in position. This brings us to the notion of an ordering. An ordering is simply a way of arranging things into a sequence. For example, suppose we give you ten T-shirts of varying shades of blue and tell you to arrange them from the darkest blue to the lightest blue. If you do so, then you have produced an ordering of those shirts. But this is not a *temporal* ordering. So some orderings are non-temporal.

What would it take to produce a temporal ordering? At a minimum, an ordering is a temporal ordering only if it is an ordering of times. However, merely arranging times in some sequence or other is not sufficient to produce a temporal ordering. For example, suppose you arrange all of the instants of the universe in terms of which instants are your favourite moments. Thus, you start with all the instants containing your birthdays since you like those best. Then you include the instants that correspond to the existence of dinosaurs, since you're rather fond of those too. Lastly, you include the instants that correspond to you going to the dentist, since those are your least favourite. This is not a temporal ordering, since there is no sense in which your ordering of times reflects any temporal facts about the universe. Indeed, your ordering seems to *jumble* the temporal facts. It places dinosaurs after your birthdays, when in fact not a single one of your birthdays is before a moment containing a dinosaur on any reasonable temporal ordering.

So some orderings of times are temporal orderings and some orderings of times are not temporal orderings. What, then, makes an ordering a temporal one? Here's one natural answer: what makes an ordering of times temporal is that the ordering has an in-built direction, and there is some fact of the matter as to which time is really present (and hence which past, and which future) and that which time is present, changes. Here is the idea. Consider Sydney and Singapore. We can order these places, and there is, clearly, a distance between them. Yet we are not inclined to think that the ordering relation that holds between them is temporal. Why not? Well, in part because we can go from Sydney to Singapore, or from Singapore to Sydney: there is no direction to the relation that holds between them. It's not only possible to travel both ways, but there's no sense in which if you travel from Sydney to Singapore you're really going in the wrong direction! So the relation that holds between them is *undirected*. But that seems very unlike our intuitive view of time. Time seems to be directed: it seems as though time goes from the past to the future rather than the other way around. More than that, it seems as though which moment of time is present, *changes*. It's not only that some time in 1930 was once present, and has stayed that way for evermore: instead, some time in 1930 *was* present, and then *became* past, and now, some time in 2018, is present.

It was McTaggart who first suggested that in order for an ordering of times to be temporal, it has to be that the ordering has a direction. Moreover, he thought, the only way for there to be a direction is for one of those times to be really present, and for which time that is, to change. Then the direction of time is given by the movement of presentness: by the fact that, first, earlier times in the ordering are present, and later, later times are present. That's what determines that time goes from the past, to the future: from earlier, to later. This way of ordering times is the A-series.

The A-series, more precisely, is an ordering of times in terms of intrinsic, monadic (i.e. non-relational), irreducible properties of *being present, being past* and *being future*. These three notions are taken to correspond to distinct metaphysical features of the universe. The present is some metaphysically special moment, and the past and future are defined in terms of their relationship to the present. In an A-series ordering there is exactly one time that possesses the property of being present. For any time that is earlier than the present, that time possesses the property of being past, and for any time that is later than the present, that time possesses the property of being future. Crucially, *which* moment is present, changes, so the A-series ordering itself changes. Which time is present shifts from moment to moment. So, for instance, on Tuesday at 2pm, 2pm is the present moment. This means that anything before 2pm is past and anything after 2pm is in the future. The present, however, shifts from 2pm to 3pm with the passage of time. When 3pm is present, 2pm will be past and everything after 3pm will be future. And so on. Time is constantly in flux: the present keeps moving from moment to moment. This change in which moment is present is the promised sharpening of temporal passage, because the present itself *moves*, and hence time itself passes.

Not everyone agrees that for an ordering to be a temporal ordering it must be an A-series ordering of times. In addition to the A-series, McTaggart identifies two further temporal orderings: the B-series and the C-series. The B-series orders times in terms of what are known as B-relations. These relations are the relations of *earlier-than*, *later-than* and *simultaneous-with*. The B-relations of earlier-than and later-than are asymmetric. If X is earlier than Y, then it is not the case that Y is earlier than X. B-relations are also unchanging: if X is earlier than Y, then this is a fact about the universe that never changes. Importantly, the B-series does not include any facts about whether events are *really* present, past or future, and makes no use of monadic properties that correspond to these three temporal attributions. Nonetheless, B-relations are, also, directed: there is a fact of the matter as to which direction is past, and which future: time goes *from* earlier, *to* later (exactly what is responsible for generating this direction is controversial, and something we will consider in Chapter 5).

The C-series – which is the most minimal of the three temporal orderings – is one that orders time in terms of the C-relations. McTaggart thought of C-relations as being non-temporal, but, as we will see much later on, that is controversial. The best way to think of the C-relations is as *undirected* asymmetric ordering relations such as *between*, so that we can say things like: the Second World War is between the big bang and the Vietnam War. Moreover, these relations are unchanging: if X is between Y and Z, then this is a fact about the universe that never changes. Finally, there is no sense in which any time in this ordering is really present, past or future. In short, C-relations share much in common with B-relations. However, C-relations are more like relations of *greater-than* and *less-than*. The greater-than and less-than relations allow us to order the natural numbers from 0 to 10. And these relations are asymmetric: if X is greater-than Y, then Y is not greater-than X. But these relations are undirected in the following sense: the natural numbers don't *really* run from 0, to 10 (or from 10, to 0). There is no fact of the matter as to direction of the number sequence.

To summarise the discussion so far: we started with the question of what makes an ordering of times a *temporal* one. We have now discerned three answers to this question:

[A-series] An ordering of times is temporal if it is (i) directed (ii) changing and (iii) orders times in terms of real properties of being present, past and future.

[B-series] An ordering of times is temporal if it is (i) directed (ii) asymmetric (iii) unchanging and (iv) orders times in terms of relations of earlier-than, later-than and simultaneous-with.

[C-series] An ordering of times is temporal if it is (i) asymmetric (ii) unchanging and (iii) orders times in terms of a betweenness relation.

We can use these three series to flesh out the distinction between the dynamic and static theories of time introduced above. Let us start with the dynamic theory of time. The dynamic theory of time takes the A-series very seriously. According to all dynamic theories of time, there is a genuine A-series – that is, an A-series that is not merely perspectival (i.e. some feature of our psychology rather than a feature of the world, more on this later on), but where presentness (and perhaps also pastness and futurity) are real features of the world. Thus, we can now redefine the dynamic theory of time as the A-theory of time. According to this view in its most minimal guise:

A-theory: The A-series exists.

Note that McTaggart believed that it was *necessary* for the existence of time that the A-series exists. Not all contemporary A-theorists hold such a strong view, and so we have formulated the theory as an extensional theory only. Note also that A-theorists typically believe in the existence of the B-series as well as the existence of the A-series. But, again, not every A-theorist holds such a view, and so we have formulated the A-theory in a less committal fashion.

Next: the static theory of time. Using McTaggart's distinction between the B-series and the C-series we can discern two distinct theories of time. First, the B-theory of time. According to the B-theory of time the A-series does not exist. However, the B-series exists. Time is static in the following sense: the relations that events/times stand in (the B-relations) are unchanging. Thus if P is earlier than Q, then P is *always* earlier than Q. Contrast this with a property like 'being present'. If P is present today, it won't be present tomorrow; but Q will be!

B-theory: (i) the B-series exists; (ii) the A-series does not exist.

The second static theory of time is the C-theory of time. According to the C-theory, the C-series exists but the B-series and the A-series do not. Like the B-theorist, the C-theorist maintains that the C-series relations are unchanging.

> **C-theory:** (i) the C-series exists; (ii) the A-series does not exist; (iii) the B-series does not exist.

What C-theorists and B-theorists disagree about is whether time has a direction: the B-theorist says it does, the C-theorist says it does not. This difference stems from the underlying difference between the two series: the B-series is directed, whereas the C-series is not. This will become important later on when we consider the direction of time in more detail.

Why would anyone believe the static theory of time? The answer, in brief, is that there are compelling arguments against the dynamic theory of time. One of these is a scientific argument: the static theory of time appears to be implied by our current best physics. Thus our experience of the world seems to pull us in one direction, and our best science in another. We are familiar with these sorts of conflicts from the history of science. Consider, for instance, the view that the Earth is flat. Our everyday experience of the world would seem to suggest that the Earth is flat: the Earth certainly *looks* flat. Our best science, however, tells us that the Earth is a sphere (or near enough). Similarly, our experience would seem to suggest that time passes. Our best science, however, strongly suggests that it does not.

Clearly the scientific case against the flat Earth is conclusive. But what about the case against the dynamic theory of time? Before we can consider the case against the dynamic theory and in favour of the static theory, we need to get a bit clearer on the two theories. We also need to spend a bit of time considering the arguments for and against the dynamic theory so that we can see just how serious the conflict with science turns out to be. In what remains of this chapter, we will complete this first task and explain how to conceptualise these two important theories of time. In the following chapters, the case for and against the dynamic theory will be considered in greater detail.

1.3. Temporal Ontology

So far we have seen how to cast the basic distinction between the static and dynamic theories of time in terms of McTaggart's three-way distinction between the A-series, the B-series and the C-series. This is useful as it allows us to get beyond metaphorical characterisations of the distinction to something more precise. But McTaggart's framework does not quite capture everything that we want to say about time, and so it leaves important questions regarding the static and dynamic theories of time unanswered. In particular, it fails to capture an important *ontological* question, a question about what *exists*.

Consider yourself as you sit there reading this book. Now consider Dino, the dinosaur, and Marvin, the intelligent robot who will be built in the year 2080. Of these three beings, who exists? Unless you are very sceptical indeed you are probably pretty sure that you exist. But what of Dino and Marvin? First, notice that the question of whether Dino and Marvin exist had better not be the question of whether Dino and Marvin exist *now*, that is, at this moment. Presumably that question is easy to answer: no, clearly they do not. That, however, is not the sense of existence at issue. To see why, suppose we ask you whether Singapore exists. Let's suppose that we are not in Singapore and neither are you. When we ask you the question, we are not asking you whether Singapore exists *here*. Clearly Singapore does not exist *here*, since none of us is in Singapore. Yet we think Singapore exists all the same, because when we ask the question 'does Singapore exist?', what we mean to ask is whether, in the broadest possible sense of 'exists', Singapore exists. Does Singapore exist somewhere in space? Assuming you are not a Singapore sceptic, we assume you think that Singapore does exist.

But what of Dino and of Marvin? Just as we can ask whether Singapore exists somewhere we can equally ask whether Dino and Marvin exist *somewhen*. That is, just as we might wonder whether Singapore is out there existing in space somewhere, so too might we wonder whether Dino and Marvin are out there existing in time *somewhen*. There, the answer to the question is much more controversial. We can distinguish three answers to the question of which concrete objects and events exist (see below). Note that by 'exists *simpliciter*' we have in mind this broader sense of existence; the same broader sense of existence used when we ask after Singapore's existence.

ENTITY NOWISM: Only entities that exist now, exist *simpliciter*.

ENTITY NOW AND THENISM: Only entities that exist now, and exist in the past, exist *simpliciter*.

ENTITY EVERYWHENISM: Entities that exist now, exist in the past, and exist in the future, exist *simpliciter*.

According to ENTITY NOWISM, you exist and we exist, but Dino and Marvin do not. According to ENTITY NOW AND THENISM, you exist and we exist, as does Dino, but Marvin does not exist. According to EVERYWHENISM, you exist, we exist, Dino exists, and Marvin exists. We all exist at different places in time, of course, just as Sydney exists at a different spatial location from Singapore. Nevertheless, Dino and Marvin are as real as you are, just as Singapore is as real as Sydney and Zimbabwe.

There's something deeply intuitive about ENTITY NOWISM. It's odd to think that Dino is out there somewhen, as real as you are. It's even odder to think that Marvin is out there somewhen. After all, the person who designed Marvin hasn't even been born yet (though they too exist in the future somewhen). Given that this person hasn't been born yet, it seems as though that person could still decide not to build an intelligent robot. Yet if Marvin exists somewhen, it seems as though his designer is somehow destined to design and build him.

There's also something intuitive about ENTITY NOW AND THENISM. For notice that we think of the past as over; done and dusted; been and gone, fixed and immutable. If it was the case that, say, Dino had twenty-two feathers on his head, then that will always be true: it cannot, now, be changed. By contrast, it seems as though the future is open to our manipulation. It seems as though how many hairs Marvin has on his head (if any) is something that is still to be determined. Since ENTITY NOW AND THENISM supposes past objects to exist, but supposes future objects not to exist, it has a ready explanation for why the past seems different in these ways from the future. For Dino exists, in time, and however many hairs are on his head determines its now being true, or false, that 'Dino has twenty-two hairs on his head'. But Marvin does not yet exist, and so it remains, as yet, unsettled how many hairs will be on his head. It is the existence of past objects that explains why the past is fixed, and why there are truths about the past, and it is the non-existence of future objects that explains why the future is malleable, and why there are (at least sometimes) no truths about future objects.

ENTITY EVERYWHENISM is the view that Dino, you, and Marvin, exist. All of these objects exists somewhere in the totality of spacetime, though of course they are separated by spatial and temporal distances, just as Sydney is spatially separated from Singapore. It might initially seem that

ENTITY EVERYWHENISM is the worst of all three options, since it is committed both to the existence of Dino and of Marvin. So it cannot appeal to any difference in their ontological status to explain why it seems to us as though facts about Dino are fixed and unchangeable, but facts about Marvin are not. As we will see, however, ENTITY EVERYWHENISM is a popular view amongst philosophers. In part (though only in part) this is because, as we will see as we explore the physics of time, the theories of general and special relativity give us reason to think that ENTITY EVERYWHENISM is true. Equally, certain philosophical puzzles arise if we accept ENTITY NOWISM or ENTITY NOW AND THENISM – puzzles we will talk about in the following chapters – and these puzzles have led some (but by no means all) philosophers to embrace ENTITY EVERYWHENISM.

For now, the important point is simply that questions about what exists and *when* can have a range of different answers, and intuitions might drive us to prefer some of these answers to others. In what follows we will inject the threefold distinction between NOWISM, NOW AND THENISM, and EVERYWHENISM back into the distinction between the sharpened versions of the static and dynamic theories of time outlined above. This will yield a richer understanding of those views, and a broader space of theoretical options for understanding time.

1.4. The Block Universe

Let us begin with the static theory of time. By far the most influential static theory of time is what is often known as the block universe theory, and what we will call the standard block universe theory. The standard block universe theory combines ENTITY EVERYWHENISM with the B-theory of time. Remember, the B-theory of time says that the B-series exists, and the A-series does not exist. So the standard block universe theory, which we can define simply below, is the view that the B-series exists, the A-series does not exist, and all concrete entities past, present, and future, exist *simpliciter*.

> **Standard Block Universe Theory:** B-theory + ENTITY EVERYWHENISM.

We can contrast the standard block universe theory with what we call the minimal block universe theory, which is very much like the standard theory, except that rather than endorsing the B-theory, it instead endorses the C-theory. So the crucial difference between the two views lies in what they think characterises temporal relations: namely, in whether or not they think that temporal relations have a direction and hence, time itself has a direction (standard theory), or whether they deny this claim (minimal theory).

Minimal Block Universe Theory: C-theory + ENTITY EVERYWHENISM.

All versions of the static theory of time deny that events undergo any kind of meaningful shift from being future to being present to being past. Time does not literally pass in this fashion. Rather, according to the static theory of time, all of the events that ever were, are, or will be, are all 'already out there', in the universe. We, as observers, stumble across these events as we 'move' through space and time.

The block universe theory in both its guises might seem a bit perplexing. On the one hand, it says that the A-series does not exist. But, on the other hand, it talks about past, present and future. So what gives? According to the block universe theory, terms such as 'past', 'present' and 'future' are to be understood in terms of either the C-relations (if one is a minimal block theorist) or the B-relations (if one is a standard block theorist). Since the B-theory is the more common version of a static theory of time, in what follows we will spell out these notions as the B-theorist would. But it is worth noting that the C-theorist can also provide an account of what we mean by these terms, it is just that her account will appeal to C-relations plus certain features of our local environmental and psychological orientation.

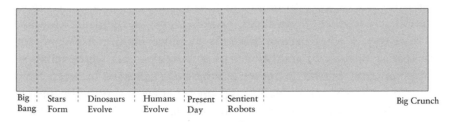

| Big Bang | Stars Form | Dinosaurs Evolve | Humans Evolve | Present Day | Sentient Robots | Big Crunch |

Figure 1 The Block Universe: This figure represents the ontology of both versions of the block theory. Note that the entire extent of time exists, from the big bang to the big crunch.

So, for example, suppose that at 10pm on Tuesday evening, Sara is dividing the events in her life into past, present and future events. According to the B-theorist, what it means for an event E to be present for Sara is just for E to be simultaneous with Sara's experiences at 10pm on Tuesday evening. Similarly, what it means for E to be past, is just for E to be earlier than those experiences. And what it means for E to be future is just for E to be later than those experiences. In short, 'past', 'present' and 'future' are always to be understood as being relative to a particular perspective in time. Thus, no event is ever *just* past, according to the static theory of time. Rather, events are past *relative to* some particular location in the universe. What is past relative to one location may well be future relative to other locations.

In this manner, temporal notions such as 'past', 'present' and 'future' are very much like spatial notions, such as 'here' and 'there'. When we say that there is a cyclone here, what we mean is that relative to the particular spatial location at which we happen to be located, there is a cyclone. Accordingly, an object or an event is never *just* here or *just* there. Rather, events and objects are here or there relative to a particular spatial location. Here and there are always relative to a position in space. In much the same way, past, present and future are relative to a position in time.

The analogy between space and time in the static theory is tight, but it is important not to overemphasise the analogy. Despite being perhaps more similar than we might have initially thought, space and time remain distinct. Exactly what the difference is between space and time within the static theory is an interesting, and vexing, question. One natural thought, however, is that the difference between time and space has something to do with causation. If we focus on a particular spatial location over time, then we can see that events located within that spatial location will cause other events to occur. So, for instance, if we consider a particular piece of sheet ice in the Antarctic located in a particular region, we can see that over time sun causes the ice to melt. But now consider a particular temporal location across space. If we focus, say, on 10am on Tuesday 27th March 2017, we see a world frozen in space. As we move, in our mind's eye, from one spatial location to another across this single time, we certainly see spatial variation, but we do not see anything like causation between events occurring, whereby one event brings some other event about. The universe is rather like a photograph, frozen mid-action.

Time, and not space, it would seem, is the dimension across which causation occurs. There is a lot more to say about this way of cleaving time from space and, later on in the book, we will have reason to revisit the relationship between time and causation. But what we have said so far is sufficient to give a basic understanding of the difference between time and space within a static theory.

Before going any further it is important to head off a common misconception about the static theory of time. The word 'static' can make it seem as though everything is frozen in place; nothing ever changes. But the static theory of time is *not* the view that nothing ever changes. Rather, the static theory of time denies only the existence of a certain kind of change, namely the change in which things possess which A-properties as time passes.

But then what *is* change according to the static theory of time? The static theory of time appeals to an 'at-at' notion of change. What this means, roughly, is that change is to be analysed in terms of objects or events having one property at a time t_1 and a distinct property at a time t_2. So, for instance, suppose that Sara is sick on Monday and then well on Tuesday. Sara changes, according to the at-at theory of change: she has changed from being sick to being well. This change is underwritten by the possession, on Monday, of the property 'being sick' by Sara, and then her possession, on Tuesday, of the property 'being well'.

Some philosophers find the at-at theory of change unsatisfying. One source of this dissatisfaction concerns the apparent similarity between change over time and spatial variation. Consider a metal bar that is partially resting in the fire. This metal bar has different properties at different spatial locations. The end that lies in the fire has the property 'being hot' at spatial location s_1. The end that lies outside of the fire has the property 'being cool' at location s_2. Despite there being spatial variation in the temperature properties of the rod, it seems wrong to say that the rod *changes*. Rather, the rod simply varies across space, but it is not changing in any meaningful sense.

The worry, then, is that the at-at theory is not really an account of change. At best it is an account of something more like 'variation' across time. *Real* change across time requires something else. For the dynamic theorist, that 'something else' is the passage of time. We will not try to adjudicate on this issue here. Suffice it to say that our current best physics treats change in precisely the way that the static theory of time treats change, in terms of the having of different properties at different times. Even though physics makes no use of temporal passage as a basis for change, it seems to be very successful at modelling change nonetheless. This casts doubt on the dynamic theorist's concern that 'real change' requires passage.

1.5. The Dynamic Theories of Time

We return now to the dynamic theory of time. Really, though, we should say dynamic *theories* of time, as there are many. Each of these theories aims to offer a model of time in which temporal passage is a real feature of reality. In this section we outline the three primary dynamic theories of time. These are the theories that have received the most attention in the philosophy of time since McTaggart drew the influential distinction between the A, B and C-series. After we have outlined these theories, we will explain how one might go about selecting one dynamic theory to be the best dynamic theory of time. This should be seen as more of a demonstration of how arguments in this area proceed, and should not be construed as the final truth of the matter. Part of your job as a budding philosopher of time is to come to your own view about which theory of time is best.

Remember, dynamic theories of time maintain that temporal passage is a real, mind-independent feature of reality. The disagreement between the dynamic theorist and the static theorist can be stated in terms of the temporal concepts of 'past', 'present' and 'future'. As we saw, for the static theorist, these concepts are understood as perspectival notions. What is past, present or future for an observer depends on that observer's relative location in time. For the dynamic theorist, 'past', 'present' and 'future' are not perspectival notions. Rather, these three notions are taken to correspond to distinct metaphysical features of the universe, encoded by the A-series. The dynamic theories of time differ with respect to their ontological attitudes towards past, present and future entities. Accordingly, as with the static theory of time, the dynamic theories of time can be defined in terms of the three-fold distinction between ENTITY NOWISM, ENTITY NOW AND THENISM and ENTITY EVERYWHENISM.

1.5.1. Presentism

Our first dynamic theory of time – presentism – is the view that combines the A-theory with ENTITY NOWISM. So presentists hold that only present concrete entities and events exist. So while it is the case that the USA exists because it is in the present, dinosaurs do not exist, since they lie in the deep past. Similarly, while it is the case that Australia exists, because it is in the present, the colony on the moon does not exist, since it lies in the future. In short, presentism has a very thrifty ontology. Compare this with the full-blown ontology of the block universe theory, which accepts ENTITY

EVERYWHENISM, and hence holds that dinosaurs, moon colonies and much more besides exist.

While the presentist maintains that only present concrete entities and events exist, which entities exist, changes. So, for instance, although the moon colony does not exist, it *will* exist as the passage of time sweeps forwards. Similarly, although dinosaurs do not exist, they *used* to exist, but have been thrown into nothingness. For the presentist, then, the passage of time involves the progressive coming into existence of new entities and new events as the present moves into the future, and the going out of existence of entities and events as those events cease to be present.

Presentism: A-theory + ENTITY NOWISM.

The presentist view, then, is one where reality is characterised by a continual process of creation and destruction as new entities and events come into being as time passes, and previously present entities cease to be (see Figure 2).

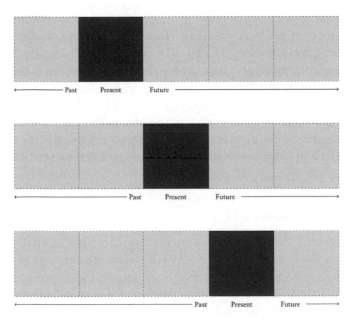

Figure 2 Presentism: Only present entities exist. Past and future entities do not exist. Which present entities exist changes with the passage of time. Notice, this diagram might make it seem as though according to presentists, the present has a temporal width (i.e. a duration). Most presentists take the present to be an instantaneous slice of reality.

1.5.2. Growing Block

The second dynamic theory of time we will consider is the growing block theory. The growing block theory combines the A-theory with a different ontological view: ENTITY NOW AND THENISM. So the growing block theorist, like the presentist, denies the reality of future concrete entities and events. The growing block theorist and the presentist also agree about the present: present entities and events exist. The growing block theorist and the presentist disagree, however, over the existence of past concrete entities. According to the growing block theorist, past entities and events exist. Thus, while the growing block theorist denies that the first moon colony exists, they accept that dinosaurs exist 'out there' in time. Like the presentist, however, the growing block theorist maintains that the first moon colony *will* exist, as the passage of time sweeps forwards.

> **Growing Block:** A-theory + ENTITY NOW AND THENISM.

By endorsing the A-theory, the growing block theorist gets to say that which moment is present, changes. How so? According to the growing block theorist, to be present is to be the latest addition to reality. The most recent entities to come into existence are the present entities. If we imagine all of the entities that exist in the past as forming a large block (much like a block universe) where that block is constituted by a stack of slices, the present entities will be the last slice in the block. The flow of time for the growing block theorist involves the coming into existence of new entities which then move from being present to being past. A representation of the ontology of the growing block theory is to be found in Figure 3.

1.5.3. Moving Spotlight

The next dynamic theory of time is the moving spotlight theory. This theory incorporates the A-theory with the very same concrete ontology as the block universe theory. So it is the view that incorporates the A-theory alongside ENTITY EVERYWHENISM. Moving spotlight theorists thus agree with the growing block theorist that past and present concrete entities exist. But they disagree with both the growing block theorist and the presentist since they hold that future concrete entities exist. Since the moving spotlight theory accepts the same concrete ontology as the block universe theory, we might wonder what the difference is between the two views.

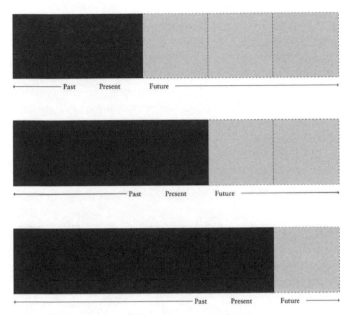

Figure 3 Growing Block: Past and present entities exist. Future entities do not exist. The sum total of reality grows with the passage of time. As with presentists, growing block theorists typically think that the present moment is an instantaneous slice of reality.

To answer this question, we need to consider the moving spotlight theorist's account of temporal passage. For the moving spotlight theorist, temporal passage is to be understood in terms of properties. The moving spotlight theorist maintains that there are special temporal properties that entities and events possess, namely the properties of being present, being past and being future. These, it will be recalled, are the properties that constitute the A-series. Furthermore, every entity or event in time possesses one of these properties. So, for instance, the USA has the property 'being present', because the USA is a present part of the world. Dinosaurs, however, despite existing in the past, lack the property of 'being present'. Rather, dinosaurs possess a distinct property, the property of 'being past'. Similarly, the first colony on the moon possess the property 'being future'. These properties underwrite the passage of time. The passage of time, for the moving spotlight theorist, involves the progressive shift in which events possess which temporal properties. Specifically, as time passes events shift from possessing the property 'being future' to possessing the property

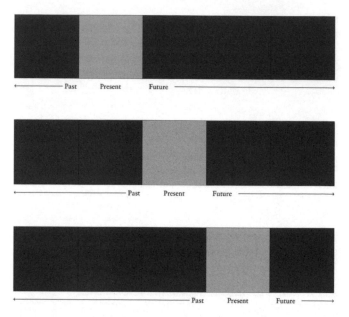

Figure 4 Moving Spotlight: Past, present and future entities exist. The present sweeps across reality with the passage of time. (Again, for ease the present is **represented here** as though it has temporal duration; but moving spotlight theorists think the present is instantaneous.)

'being present' and then, finally, possessing the property 'being past' (see Figure 4).

The difference, then, between the moving spotlight theory and the block universe theory of time concerns the existence of these temporal properties. The block universe theorist does not posit the existence of properties of 'being present', 'being past' and 'being future', at least not of the kind that the moving spotlight theorist believes in. The moving spotlight theorist's temporal properties are not perspectival: their existence is a mind-independent fact about the world.

Moving Spotlight: A-theory + ENTITY EVERYWHENISM.

1.6. How to Select a Dynamic Theory

So far in this section we have looked at a range of dynamic theories of time. But which one should we select, and why? In this section we will briefly outline an argument in favour of the view that *presentism* is the best dynamic theory of time available. We should emphasise once again, however, that this is simply a demonstration of how one might argue for a particular dynamic theory of time.

Consider the moving spotlight and growing block theories of time. According to both of these views, there is a single objectively present set of entities and events. There are also entities and events that are non-present, either by being objectively past (in the case of both views) or by being objectively future (in the case of the moving spotlight view).

According to either theory, non-present events, people and so on are just as real as present events, people and so on. Now, consider Sara as she sits in front of her pancakes on a Tuesday morning. As she is sitting there, Sara reflects on her own experiences of the wonderful fluffy looking pancakes and forms the following belief: I am present. That is, Sara believes herself to be located in the present moment. Now, consider someone who is located in the deep past: the philosopher Hypatia. As Hypatia is working through a mathematical proof on an ancient Tuesday morning, she reflects on her own experiences of the proof she is developing and forms the following belief: I am present. That is, Hypatia believes herself to be located in the present moment.

Compare Sara's and Hypatia's beliefs. At most, one of Sara or Hypatia can be right about their beliefs: only one of them believes truly that they are located in the present. And yet, their epistemic situations seem to be identical in all relevant respects. They are both fully aware of their senses, they are both having experiences, they are both forming beliefs on the basis of those experiences in a responsible manner. But now consider, again, Sara's beliefs: for all she knows, she is located in the deep past. For all she knows, her beliefs about the present are just like Hypatia's: completely wrong. It would seem, then, that Sara and Hypatia are thrust into a sceptical scenario: neither of them can be sure that they are located in the present.

Actually, we can make things a little bit worse for both Sara and Hypatia. There is only one time that is present, according to both the growing block and moving spotlight theories. But there are many times that are non-present. Indeed, there are *many more* non-present times than there are present times. So if Sara or Hypatia had to make a bet on whether or not they are located in the present or the non-present, each should bet on being located somewhere non-present, since that is overwhelmingly the most likely outcome. So each should rationally believe that she is located somewhere that is non-present (of course, rationally believing something is not the same as knowing it: one can rationally believe that one is a frog if presented with erroneous evidence of a certain kind that one has reason to trust, even though it is false that one is a frog and so one cannot know that one is a frog, because one cannot know a false thing).

Why is this a problem? Well, remember that the dynamic theory of time is motivated by the idea that we experience temporal passage. As we have also seen, the passage of time is understood in terms of a moving present: there is a metaphysically special moment – the now – and which moment that is changes. In order to experience the passage of time, then, one needs to be experiencing the movement of the present. This would seem to require that one is located in the present and is carried forwards by the motion of the present, which is then fed back into experience. But now consider Sara and Hypatia. If Sara should rationally conclude that she is in the past, then she should rationally conclude that she is not experiencing the passage of time. Indeed, each of us should reach exactly the same conclusion. But if that's right, then we have no basis to argue in favour of the dynamic theory of time. We should all believe that we *lack* the very experiences of the passage of time that are supposed to motivate the dynamic theory. In short, both the growing block and moving spotlight theories of time seem to place us all in a sceptical scenario that undermines the basis we have to believe those theories in the first place!

Presentism avoids the worry. According to presentism, only present entities and events exist. But only existing people can form beliefs and have experiences. So it follows that only existing people can experience the world and form the belief that 'I am present'. Which means that only present people can have those experiences. So if presentism is true, one can be sure that one is present, merely in virtue of having experiences and forming beliefs of any kind. So, one might conclude, if the A-theory is true, then presentism is true, since it is the only version of the A-theory that avoids these sceptical worries.

Presentists avoid the problem above by radically shrinking their ontology. By reducing the number of entities that exist, the presentist avoids imbuing past and future people with full-blown conscious experiences that are just like ours. But in reducing their ontology in this fashion, they then open themselves up to a host of new worries. The most significant problem for presentism is the so-called truthmaker objection. The objection takes many forms, but can be stated as an inconsistent triad:

[1] Presentism is true
[2] Some claims about the past are true
[3] The truth of a claim depends on being

The apparent inconsistency of [1]–[3] is not difficult to reveal. Suppose that presentism is true. If presentism is true, only present entities exist. Now, suppose that some claims about the past are true as well. Suppose, for instance, that the claim 'Hypatia was wise' is true. The truth of a claim depends upon being. One way to understand this idea is as follows: the truth of a claim depends on being when there exists some entity in virtue of which the claim is true. So, with respect to Hypatia, what this means is that if 'Hypatia was wise' is true, this implies that there exists some entity that makes this claim true. It would seem, however, that the entity in question should be Hypatia: the reason why 'Hypatia was wise' is true is that there exists an entity – Hypatia – who is in the past and who is wise. But then it would seem that a past entity exists, which contravenes presentism!

Suppose the argument from truthmaking succeeds. What should we conclude? Well, if presentism is the only reasonable version of the A-theory (as the argument above suggested) and if this argument succeeds, then it follows that there is no good version of the A-theory. Conversely, if one is a dynamic theorist of time, then it would seem one's job description is pretty clear: defend presentism from the truthmaker objection and, more generally, from any problem that presentism might face. This tends to make presentism an important dynamic theory of time to focus on.

1.7. For the Advanced Student

To summarise the discussion so far: we began with an intuitive distinction between two ways of thinking about time: the static approach and the dynamic approach. We then sharpened these approaches into theories of time using McTaggart's three-way distinction between the A-series, the B-series and the C-series. Finally, we used a further ontological distinction to provide further development of both static and dynamic theories of time.

What we have said so far provides enough background to start digging into the rest of the book. So if you are satisfied with your understanding of the static and dynamic theories, then you can move on to Chapter 2. If, however, you desire a bit more detail and want to see an even greater number of theoretical options for thinking about time, then we recommend reading on. For in this section, we will add further layers of complexity onto the basic foundations laid out here. Doing so is useful for navigating the modern literature on the static and dynamic theories of time. If you venture beyond the pages of this book and into philosophical discussions of time, you will see a plethora of other dynamic and static views available. Our aim is to give you a bit of background to understand some of this more challenging material. Here's what we'll do. First, we'll draw a few more distinctions. Then we'll use these distinctions to produce a taxonomic overview of a range of theories of time.

1.7.1. What are Times?

When we introduced the distinction between the A-series, the B-series and the C-series, we introduced the notion of *a time*. We moved rather quickly over the question of what a time is, so that we could make some headway in understanding the static and dynamic theories of time. We said simply that a time is a three-dimensional 'slice' of the universe: the entire universe in space at an instant. Let us now pause for a moment to think a bit harder about what a time *is*.

There are two major views about times. According to the first, times are concrete objects, and according to the second, they are abstract objects. Concrete things are things that are located in the physical universe and have causal powers. They are contrasted with abstract things such as numbers, sets, or functions, which are not located in space or time and lack causal powers. Most ordinary objects – dogs, toasters, quarks, electrons, galaxies and paintings – are concrete.

According to the first view of what times are, which we will call CONCRETISM, times are *maximal instantaneous concrete* entities. Imagine our *whole* universe at an instant – imagine where all the objects are, what properties they have, and how they are distributed. To say that a time is *maximal* is to say that it is *all of the* universe at an instant and not just some part of the universe at that instant. Since our world has three spatial dimensions the object we have just described is a huge three-dimensional object: it stretches a long way (perhaps infinitely) along the spatial dimensions, but it has no temporal width. According to CONCRETISM, that giant three-dimensional object is a time.

CONCRETISM: Times are maximal instantaneous concrete entities.

The opposing view is that times are abstract objects. To get a feel for this, recall that we earlier introduced the idea of possible worlds. This is the entire way that the world might be: a possibility. Now, you might think that there aren't literally any such worlds. There is just our world. You might then be inclined to think that a possible world is a kind of representation: a way of imagining a possibility, like a *story* about how the world might have been. A possible world (thought of as an abstract object) is a representation of an entire world in roughly this sense. By contrast, times are abstract representations of *instantaneous* states of the world. So where the CONCRETIST says that times are maximal instantaneous concrete entities, ABSTRACTISM says that times are *representations* of these maximal concrete entities. Again, think of a story: just as you might tell a story about how the world might have been at some time, so too might you tell a story about what a particular moment in history was like. To say that a time is abstract is, roughly, to say that a time is a kind of representation of history in basically this sense.

> ABSTRACTISM: Times are abstract objects each of which represents the entirety of our universe at an instant.

CONCRETISM has a number of advantages over ABSTRACTISM. First, if times are just big concrete objects, then they're not different in kind from ordinary concrete objects. We can study them using ordinary scientific methodology. Moreover, as we will see in a moment, CONCRETISM has the apparent advantage that it preserves a very neat connection between which times exist and which concrete objects exist. Since times just are big aggregations of concrete objects (and relations), it follows that if past *objects* exist, so do past *times*, and if future *objects* exist, then so do future *times* (and vice versa). That makes life simple.

The flip side of the coin, however, is that this simplicity comes with a cost. For one might want which concrete *objects* exist to come apart from which *times* exist. Why? Well, you might think it intuitive that ENTITY NOWISM is true, because it seems to you very implausible that Dino and Marvin are out there, somewhen. Still, you might think that there are truths about what Dino was like (e.g. about how many feathers he had on his head). If times are abstract objects then one can say that although Dino does not exist, nevertheless past *times* exist in at least some sense. One of those past times represents Dino, and the number of feathers on his head. It is because there are these past times that the past is fixed and immutable and there are facts of the matter as to how many feathers Dino had on his head. One might even think that although past times exist, future times do not. So there is no time that represents Marvin and the hairs on his head.

That, in turn, can explain why it seems to be an open matter what Marvin will be like (or indeed, whether Marvin will ever come to exist at all). Pulling the existence of times apart from the existence of concrete things allows for this kind of flexibility.

1.7.2. Which Times Exist?

Remember we introduced the following three views about which concrete entities exist:

ENTITY NOWISM: Only objects and events that exist now, exist.

ENTITY NOW AND THENISM: Only objects and events that exist now, and exist in the past, exist.

ENTITY EVERYWHENISM: Objects and events that exist now, exist in the past, and exist in the future, exist.

Let's put aside for a moment the question of which concrete entities exist, and just ask which *times* exist. Of course, you might think that there has to be some connection between these two questions. If ENTITY NOW AND THENISM is true, then, one might think, surely the present time and past times must exist! Otherwise Dino would exist, but he wouldn't exist at any time! But put that thought on hold for a moment, and just entertain the question of which times exist. Then we have three possible views:

TIME NOWISM: Only the present time exists.

TIME NOW AND THENISM: Only the present time and past times exist.

TIME EVERYWHENISM: Past, present and future times exist.

We can now come back to our earlier question about the connection between the existence of times and the existence of concrete entities. If one accepts CONCRETISM about times, then there is a very simple connection between which *times* exist and which *concrete entities* exist. If only present objects exist, then it must be that only the present time exists. If present and past objects exist, then it must be that present and past times exist. If past, present and future objects exist, then past, present and future times exist. That's because according to CONCRETISM, times just are big aggregations of concrete entities. We can summarise this as follows:

CONCRETISM + ENTITY NOWISM = TIME NOWISM

CONCRETISM + ENTITY NOW AND THENISM = TIME NOW AND THENISM

CONCRETISM + ENTITY EVERYWHENISM = TIME EVERYWHENISM

If we accept ABSTRACTISM then it doesn't follow that just because past *objects* like Dino do not exist, past *times* do not exist, or that just because future *objects* like Marvin do not exist, future *times* do not exist. In fact, the abstractionist needs to distinguish between which times *exist* and which times are *concretely realised*.

To get a handle on this, imagine going to an art gallery that has an endless series of paintings on its walls. All these paintings represent something (let's suppose there is no abstract art in this gallery). As you walk around, you can ask yourselves which of the paintings represent some *real* part of the world. Very roughly, to be *concretely realised* is to be such that you represent some part of the concrete world. For a time to be concretely realised is for the *concrete* maximal instantaneous object that that time represents, to exist. Then the abstractionist can ask, of some bunch of existing times, which of those times are concretely realised. She has the following options:

REALISATION NOWISM: Only the present time is concretely realised.

REALISATION NOW AND THENISM: Only the present and past times are concretely realised.

REALISATION EVERYWHENISM: Past, present and future times are concretely realised.

A time can only be concretely realised if it exists. So REALISATION NOW AND THENISM can only be true if present and past times (and perhaps future ones) exist, and REALISATION EVERYWHENISM can only be true if all times exist. Equally, it doesn't make a lot of sense to think that every time that exists is concretely realised, since if one thought that, it seems that one might just as well accept CONCRETISM. So there are only three interesting combinations of views about which times exist, and which times are concretely realised. Those are the following:

TIME NOW AND THENISM + REALISATION NOWISM

TIME EVERYWHENISM + REALISATION NOWISM

TIME EVERYWHENISM + REALISATION NOW AND THENISM

The first of these is the view that only past and present *times* exist, but of these, only the present time is concretely realised. This view is a version of presentism known as ersatz presentism. It's easy to see why such a view might be appealing, since on that view there are facts about Dino (facts represented by past times) but there are no such facts about Marvin (since no future times exist), yet we can say that we exist, and that Dino and Marvin do not.

The second of these says that *all times* exist, but that only the present time is *realised*. This is also a version of presentism: it's a different version of ersatz presentism. On this view, future times exist. That might seem like a cost, but it's also a benefit. Suppose we think that some claims about the future are, now, true. Maybe we think it's now true that the sun will rise tomorrow, or that eventually there will be intelligent robots (even if we're not sure when, or that they will be named Marvin). If future times exist, then we can make sense of how those claims can now be true: they are made true by those future times.

Finally, the third view holds that all *times* exist, but only present and past times are concretely *realised*. This view is a version of the growing block theory. But it's a version of the view that says that future times exist. This view has the advantage that we can explain the difference between Dino and Marvin: Dino exists, but Marvin does not. But we can also explain why some future claims are already true: they are made true by future *times*, despite no future *objects* existing.

1.7.3. Static and Dynamic Theories Again

Using the above distinctions we can now define each of the views discussed in section 1.3. In order, those views can now be characterised as follows:

View	Times	Entities	Time Series	Nature of Times
Standard Block Universe	TIME EVERYWHENISM	ENTITY EVERYWHENISM	B-THEORY	CONCRETISM
Minimal Block Universe	TIME EVERYWHENISM	ENTITY EVERYWHENISM	C-THEORY	CONCRETISM
Standard Presentism	TIME NOWISM	ENTITY NOWISM	A-THEORY	CONCRETISM
Standard Growing Block	TIME NOW AND THENISM	ENTITY NOW AND THENISM	A-THEORY	CONCRETISM
Standard Moving Spotlight	TIME EVERYWHENISM	ENTITY EVERYWHENISM	A-THEORY	CONCRETISM

Looking at this list, we can see that there remain a number of views that we have not discussed. We wish only to highlight one important dynamic theory, a version of ersatz presentism.

View	Times	Entities	Time Series	Nature of Times	Realisation
Ersatz Presentism	TIME NOW AND THENISM	ENTITY NOWISM	A-THEORY	ABSTRACTISM	REALISATION NOWISM

According to ersatz presentism, past and present *times* exist, even though past objects do not. In some sense, then, the passage of time can be modelled in terms of a change in which abstract time is concretely realised. As time flows, times that were once concretely realised cease to be so, and a new time comes to be concretely realised. The crucial difference between ersatz presentism and standard presentism is that according to the former, past times exist, whereas according to the latter, they do not. Ersatz presentism is an important view in the modern literature in the philosophy of time. It seems to boast the advantages of a non-presentist view, without the drawbacks of a presentist view. We leave it as an exercise for the interested student to try and map further theories of time and consider what, if anything, they have going for them as compared to the standard offerings.

1.8. Beyond the Static and Dynamic Theories

We would like to close this chapter with a radical thought. Everything we have said so far assumes that time exists. This is a basic assumption of both the static and dynamic theories of time. But what if time doesn't exist? The implications would be startling indeed. The existence of time underwrites a range of important phenomena. It is, for instance, the very basis for historical investigation, as it provides a means of indexing and categorising events and persons. Without time, it is difficult to make sense of historical research; for if time does not exist, then there is no history to speak of. Similarly, time is an essential component of research within evolutionary biology. Time, in evolutionary terms, is the scale across which natural selection, gene mutation and reproduction occur. If time does not exist then the very concept of evolution by natural selection appears to be incoherent, relying as it does on generational development. In psychology, time underwrites the frequency and rate of neural firing, and thus of stimulus intensity. Indeed, if there is no such thing as time, then it is difficult to make sense of even a single neuronal firing, let alone an entire suite of neurons working together to produce anything as interesting as a conscious human being.

Similar examples may be found across a number of other areas: chemistry, biology, cosmology, social science, archaeology (and many more besides). All of these fields assume the existence of time, and scarcely make sense without it. But the importance of time is far greater still, for time reaches down into the very foundations of human experience. Consider, for instance, practical reasoning. When we make decisions we plan what we will do in the future – based on present desires and beliefs, including beliefs about the past. If time does not exist, then there is no such thing as the future. Accordingly, it is very difficult to see how anyone can reasonably decide to do anything in a timeless world. Similar considerations apply to moral reasoning: when we hold an individual accountable for wrongs committed, we are taking them to task for actions they have carried out in the past. But if time does not exist, then there is no such thing as the past, and so no reasonable sense in which anyone can be held morally accountable for anything. Moral nihilism – the view, roughly, that nothing is right or wrong – would soon follow.

Quite aside from any particular pattern of reasoning, human agency itself seems to be threatened by the loss of time, for without time it is very difficult to see how there could be anything like causation. What it is for an agent to cause another event to happen is for that agent to perform an act at an earlier time that has ramifications into a later time. Without time, concepts such as 'earlier than' and 'later than' do not seem to be in good standing, and so it would appear that no reasonable sense can be made of causal efficacy. But the power to perform an act is the very core of agentive behaviour. So the loss of time takes with it our most basic self-conception as agents. The question of whether or not time exists, then, is one that transcends the academy. We all take ourselves to be causally efficacious agents that act in the world in response to reasons that we take ourselves to have. Whether and how we are agents that act in this way is important to all of us, and so the loss of time threatens to completely revolutionise the way we think about our place in the world. It also promises to substantially reorient the way we think about history and evidence, the way we conduct scientific investigation and the manner in which we understand reality itself.

In later chapters we will return to this issue in order to say a bit more about what we call timeless physical and metaphysical theories. These are theories that explicitly deny the existence of time in one way or another. For now we wish only to point out that such theories exist, and represent an interesting alternative to standard theories of time.

1.9. Summary

In this chapter we have provided an introduction to a number of different theories of time. The key points to take away from this chapter are summarised as follows:

(1) A constitutive theory of time gives an account of the nature of time itself. An extensional theory of time tells us about the features that time has actually, but that may not be part of its nature.

(2) Static theories of time hold that temporal passage is not a real feature of reality; dynamic theories of time hold that temporal passage is a real feature of reality.

(3) There are two versions of the static theory of time: one that combines the B-theory with ENTITY EVERYWHENISM and one that combines the C-theory with ENTITY EVERYWHENISM.

(4) There are many dynamic theories of time. Each of these differ with respect to what they take to exist. Presentists hold that only present entities exist, growing block theorists maintain that past and present entities exist (but not future entities), and moving spotlight theorists maintain that past, present and future entities all exist.

1.10. Exercises

i. Make a table that maps all of the different dynamic theories of time.

ii. Try to invent a new dynamic theory of time that is different to any of the ones discussed here.

iii. Try to come up with an example where you would have compelling evidence that there is a period of time in which nothing changes.

iv. Try to think of two problems for each of the following views: presentism, the block universe, the growing block, the moving spotlight.

v. Consider the evidence in favour of the dynamic theory of time and the evidence in favour of the static theory of time. Try to argue that the evidence for one of these views is superior to the evidence for the other.

1.11. Glossary of Terms

Actual World
The way that our universe is.

Mind-Independent
Something is mind-independent when it does not depend for its existence on the existence of consciousness.

Necessary Condition
A is a necessary condition for B when B is required for A to be the case.

Possible World
A complete specification of a universe that is spatiotemporally disconnected from our universe. Possible worlds are ways things could be. The set of possible worlds is the set of worlds that represent the way the actual world could have been.

Sufficient Condition
A is a sufficient condition for B when A is enough for B to be the case.

Temporal Ordering
An ordering of events in time, or of instants of time. Temporal orderings are often thought to be linear total orderings.

1.12. Further Readings

K. Miller (2013) 'The Growing Block, Presentism and Eternalism' in Heather Dyke and Adrian Bardon, eds, *A Companion to the Philosophy of Time* (Wiley-Blackwell), pp. 345–64. An accessible introduction to both static and dynamic theories of time.

M. Sullivan (2012) 'Teaching & Learning Guide For: Problems with Temporary Existence in Tense Logic', *Philosophy Compass* 7 (4): 290–2. A fairly accessible overview of A-theories, alongside a consideration of tense.

C. Hare (2010) 'Realism about Tense and Perspective', *Philosophy Compass* 5 (9): 760–9. This is a more challenging overview of different ways of spelling out the A-theory.

2

The Passage of Time

In Chapter 1 we distinguished static from dynamic theories of time. According to dynamic theories of time temporal passage is an objective, mind-independent feature of reality. When we introduced the concept of temporal passage we noted that much of the discussion surrounding this topic is cloaked in metaphor. Our goal in this chapter is to try and get behind the metaphor even more, and provide a more sustained analysis of temporal flow. We will consider a range of philosophical arguments for the view that temporal passage is incoherent. These arguments suggest that there may be no consistent unpacking of the metaphorical language that is so often used to describe the flow of time. Along the way we will consider the prospects for addressing these arguments and thus rescuing the dynamic theory of time.

2.1. What is Temporal Passage?

One of the earliest accounts of temporal passage in the contemporary philosophy of time is the account proposed by McTaggart. As we discussed in Chapter 1, McTaggart draws a distinction between the A-series and the B-series. It is useful to briefly remind ourselves of the distinction. A B-series ordering is one that orders times in terms of the B-relations. These relations are the relations of *earlier-than*, *later-than* and *simultaneous-with*. Importantly, these relations are unchanging: if X is earlier than Y, then this is a fact about the universe that never changes. It is not the case, for instance, that X is earlier than Y on Tuesday, and later than Y on Wednesday. An A-series ordering of times is an ordering of times in terms of the A-properties. Whereas the B-series makes use of relations between times, the A-series makes use of intrinsic, monadic (i.e. non-relational), irreducible properties of being present, being past and being future. In an A-series ordering, there is exactly one time that possesses the property of being present. For any time that is earlier than the present, that time possess the property of being past, and for any time that is later than the present, that time possesses the property of being future.

As we noted briefly in Chapter 1, the A-series is one way to understand what it is for there to be temporal passage. But it is not the only way of understanding that notion. One of the limitations of understanding temporal passage purely in A-series terms is that temporal passage so understood presupposes the existence of properties of being present, being past and being future. Moreover, it assumes that every time that there is possesses one of these properties.

This way of thinking about temporal passage won't work for some of the dynamic theories of time discussed in Chapter 1. Consider, for instance, presentism. According to presentism, only present entities exist. Because only present entities exist, there are no past or future entities available that can possess properties like being past or being future. Indeed, about the only dynamic theory of time that neatly fits the A-series conception is the moving spotlight theory. And, indeed, the A-series does illuminate that view. We can now see what the moving spotlight posited by the moving spotlight model is: it is the property of being present, sweeping across the universe, leaving the property of being past in its wake, and progressively eating up the property of being future. But for dynamic theories of time that are not moving spotlight views, a different conception of the passage of time is required.

We will look at the precise nature of temporal passage for these other views in the next section. First, however, it is useful to have a general statement of temporal passage that will apply to all versions of the dynamic theory of time. To this end, we may characterise the passage of time in terms of a moving present. There is some time – the present – that is metaphysically special in some respect. Exactly which moment is metaphysically special in this manner changes. So, on Tuesday at 2pm, 2pm is the metaphysically special present. When it is 3pm, 3pm will be the metaphysically special present, and 2pm will lose its special status. And on it goes.

General conception of passage: There is an objective, metaphysical fact, regarding which moment is present, and which moment that is, changes.

The A-series fits under this general conception. In the A-series, the present is metaphysically special in virtue of the existence of a monadic property of presentness. This property moves in so far as the A-series is constantly changing. But while the above basic conception of temporal flow subsumes the A-series, it is broader still. For it allows for there to be ways of conceiving of temporal passage that make no use of monadic properties of presentness, pastness and futurity, as we shall now see.

2.2. Flavours of Passage

Each version of the dynamic theory of time will understand the notion of temporal passage just articulated in a different way. In particular, each version will provide a different account of what it is for there to be a metaphysically special moment; they differ over what it is to be present. As noted, a moving spotlight theory of time will maintain that the metaphysical specialness of the present consists in the possession, by some time, of a monadic, intrinsic, irreducible property of presentness. The flow of time, on this view, consists in the shift in which moment possesses this property.

A presentist, by contrast, will equate the metaphysical specialness of the present with existence. What it is to be present is to exist, according to presentism. The flow of time, on this view, involves a constant shift in what exists. At 2pm on Tuesday all of the things that are simultaneous with one another at 2pm exist. None of the 1pm things exist and none of the 3pm things exist. As time passes, however, the 2pm things will go out of existence and the 3pm things will come into existence. For the presentist, past and future are still understood in terms of the earlier- and later-than relations. To be past is to be earlier than the present and to be future is to be later than the present. One of the chief challenges facing presentism is how to give an account of cross-temporal relations such as these, relations that link things that exist with things that do not.

A growing block theorist, by contrast, equates the metaphysical specialness of the present with the notion of being the most recent moment to come into existence. The flow of time, on this view, involves the progressive coming into existence of new times. The flow of time is therefore *accretive*: it is constantly enlarging the sum total of existing things. So when 2pm on Tuesday is present, 2pm is the latest moment to have come into existence. When it is 3pm, 3pm will be the latest moment to have come into existence. 2pm, by contrast, will have shifted to being past, since it is no longer the latest moment to come into existence. The growing block theorist has an easy account of the distinction between the past and the future: what it is to be past is to be an existing thing that is earlier than the present. What it is to be future is to be a non-existent thing that is later than the present. As with the presentist, the growing block theorist may need to provide an account of how there can be relations between present existing things and future non-existent things.

We have before us three concepts of temporal passage. These different approaches to passage provide unique interpretations of the flow of time. They are all united, however, in their understanding of passage in terms of a moving present, but they differ over what it is for the present to move in this way. While there is disagreement of this kind, there is also a very general agreement about what's going on when time passes. On every one of the views just discussed, the passage of time involves a shift in *ontology*: what exists and which things have which properties. The moving spotlight view involves a shift in the things that possess the property of being present. Presentism involves a constant shift in what exists. The growing block view involves the gradual increase in what exists.

It is an interesting question as to whether the passage of time must be thought of in these terms. Do we have to think of it ontologically, in terms of some change in what exists and the properties that existing things possess? Or are there other ways to conceptualise this notion? We leave these questions to the reader.

2.3. Problems for Passage

Let us focus on the first concept of flow articulated above: the notion of a moving present. One of the core aspects of this account of temporal passage is that it seems to require a certain kind of change: either change in what exists, or change in which things possess which properties and so on. Because the passage of time involves a constant change in ontology, it is open to some fairly serious objections. In what follows we will outline the two most pressing such objections: the rate of flow argument and McTaggart's paradox. Before considering these objections in detail it is worth asking what they would show, were they to succeed. As will become apparent, if either argument succeeds then it shows that there is something internally incoherent in the very notion of temporal passage. As such, if either argument succeeds we should conclude that, of necessity, there is no temporal passage. That is, we should conclude that temporal passage is impossible. That leaves us with two options. We could reject both the constitutive and extensional A-theory in favour of the B-theory or C-theory, and conclude that temporal passage is not essential for there being time. Indeed, given the impossibility of temporal passage we will conclude that there is no world with temporal passage. Alternatively, we could accept the constitutive A-theory and reject the extensional A-theory. Then we will hold that temporal passage is essential to there being time, which means that we will conclude not only that there is actually no time, since there is no temporal passage, but, more strongly, that time is impossible because time would require there to be temporal passage, and temporal passage is impossible. It is worth bearing these options in mind as we consider each of the arguments.

2.3.1. The Rate of Flow Argument

The first of these objections can be set out as follows. Change is usually understood as variation over time: X changes when X takes on different values at different times. But how, then, do we understand the idea that time itself changes? Without invoking an extra dimension of time (more on this later on) the only option is to understand the idea that time itself changes in terms of times taking different values at different times. But this is difficult to understand. One way to draw out the problem is to think of the rate of such change.

If time passes, then it must pass at some *rate*. That's because all changes happen at some rate. For instance, when a light changes from red to green, that change can happen slowly (taking an hour) or quickly (taking a second). Either way, the change happens at some rate. The question about the flow of time, then, is this: if time passes at some rate, at what rate does it pass? A rate is a measure of some change over time. So how fast does time pass, thought of in this manner? The only reasonable answer, it would seem, is that time passes at one second per second: for every second that elapses, the present has moved one second into the future. But there are serious doubts about the idea that 'one second per second' is a coherent rate at all. Rates are usually understood in terms of the change of one quantity as a function of a change in another quantity. For instance, one might understand velocity in terms of the *spatial* distance travelled per unit of time. When we're talking about the passage of time, however, it seems we are forced to measure the rate of time's passing against itself. But that's just not a meaningful rate. So time has no rate. So it doesn't flow. So there is no such thing as temporal passage.

In order to address the rate of flow argument there are, broadly, three options available to the dynamic theorist. First, she might attempt to argue that the rate of flow specified above is a meaningful rate after all. There is nothing incoherent about the idea that time passes at one second per second, she might say. By way of analogy, suppose that Suzy and Billy both have ten dollars. Suppose that Suzy says to Billy: 'for every dollar you give me, I will give you a dollar back'. The exchange rate between Suzy and Billy, then, is one dollar per dollar. Now, to be sure, Suzy and Billy are not going to become millionaires by exploiting this particular exchange rate. Indeed, they cannot bring it about that either of them has more than ten dollars. But that doesn't make the exchange between them incoherent or otherwise meaningless. They can meaningfully exchange one dollar per dollar and can reasonably characterise their shifting financial situations in terms of this exchange rate.

One might worry about this analogy, however, for a couple of reasons. First, it is not obvious that the time and financial cases are genuinely analogous. In the time case, we are trading one second of time for one second of time. In the financial case, we are trading one of Suzy's dollars for one of Billy's dollars. So in the financial case, the rate of flow between Suzy and Billy involves two distinct quantities. In the time case, by contrast, we are exchanging only a single quantity. We can further grasp the difference by thinking in terms of other possible exchange rates between Suzy and Billy. Although Suzy and Billy are exchanging one dollar per dollar, they could exchange two dollars per dollar, or three dollars per dollar. Or some other rate entirely. In short, we can easily make sense of variations in the rate of exchange between them. In the temporal case, however, we cannot easily make sense of this. It just doesn't make sense to say that time could flow faster: we cannot say that time could flow at two seconds per second or three seconds per second. That's because all we have is a single quantity, and so any variation in the quantity that appears on one side of the rate of flow must be carried over onto the other side.

A second response to the rate of flow argument is to deny that the right characterisation of the flow of time is that it flows at one second per second. This is not the right characterisation precisely because it makes use of a single quantity only, and a single quantity cannot meaningfully vary with respect to itself. The challenge in defending this approach, however, is to find some second quantity against which the rate of flow can be meaningfully measured. One option, which we will explore below, is to invoke a second time dimension and use this as the measure of time's flow. Barring that, however, the only other thing that we could possibly use to measure the flow of time is space. It is quite unclear, however, how we might understand the rate of temporal flow as variation over space. Such an account would involve saying that time flows at, say, one second per metre. But what could that mean? And would it correspond to the notion of temporal passage as we have been describing it, as involving a moving present?

The third and perhaps most promising response is to take issue with the initial formulation of the problem. As the problem was formulated, it relied on the idea that the passage of time is a kind of change. If the passage of time is a kind of change, where by 'change' we mean what we ordinarily mean by that term, then temporal passage must be understood as a variation in one quantity as a function of another. But, the dynamic theorist might argue, this is to misunderstand the claim that temporal flow is a kind of change. Change, after all, is something that happens *in* time as, for example, when we say that the weather has turned foul. What we mean is that, over time, the weather has shifted from being nice to being nasty. Temporal passage is not something that happens *in* time, it is something that happens *to* time. Because it is something that happens to time, we should not expect it to obey the normal rules that change in time obeys. We should not, for instance, expect that temporal passage should be understood as a variation in one quantity as a function of another. Temporal passage is something else entirely. Accordingly, we are simply making a mistake when we ask after the rate of time's flow. It doesn't have a rate, because it is not change in the normal sense.

The trouble with this third type of response to the rate of flow argument is that it renders the notion of temporal passage mysterious once more. At best, the claim that temporal passage involves change is another metaphor, at least in so far as 'change' is understood in the usual sense. But then what is temporal passage? How should it be understood? The dynamic theorist will probably reply that there just is no answer to this question. The flow of time is a primitive, unanalysable aspect of their theory. This might seem untoward. But, actually, it is perfectly acceptable to take something as primitive in the context of a theory. Indeed, every theory has to take something as primitive. It has to start from somewhere. For the dynamic theory of time, then, that starting point is temporal passage.

2.3.2. McTaggart's Paradox

The second problem with the idea that time passes also aims to show that the passage of time is not a coherent notion. This problem, called McTaggart's paradox (named for the philosopher who first proposed the paradox) focuses on the three notions of past, present and future. Suppose, as the dynamic theorist does, that these three notions are not merely perspectival. Rather, the distinction between the past, present and future is a metaphysically substantive one. Now, suppose further that the passage of time involves a shift in which events or objects possess the properties of being future, being present and being past. That is, time passes as the moving spotlight theorist says it does. The paradox proceeds as follows. Consider a particular event, E. Suppose that E is future. As time passes, E will become present and then it will become past. Accordingly, E will come to possess all three of these temporal attributions. But no event can possess more than one of these attributions: an event cannot both be future and past, or future and present, or present and past. So it cannot be the case that E possesses all three temporal attributions. But then it would seem that the flow of time is not possible, since it requires of a single event that it possess incompatible temporal attributions.

The obvious response to the problem as it has just been stated is this: an event E never possesses incompatible temporal attributions *all at once*. Rather, E possesses these three attributions successively. E is *now* future, but it *will* become present and then, after that, it *will* become past. But the italicised tenses in the preceding sentence give us nine further temporal attributions to deal with. To say that E is now future is to say that E is future in the present. To say that E will become present is to say that E is present in the future, and to say that E will be past is to say that E is past in the future. Call these further temporal attributions 'second order' temporal attributions. There are nine such attributions:

(1) Present in the present
(2) Present in the past
(3) Present in the future
(4) Past in the present
(5) Past in the past
(6) Past in the future
(7) Future in the present
(8) Future in the past
(9) Future in the future

The paradox can be reformulated using these nine temporal attributions. As time passes a single event E will come to possess all nine temporal attributions. But some of these temporal attributions are incompatible. For example, an event cannot be both future in the past and present in the past. So E cannot possess all of these temporal attributions. So E cannot be subject to the passage of time.

Again, the obvious response is: E does not possess all nine of these attributions at once. Rather, it possesses these attributions successively. It *was* present in the future, it *will* be past in the present and so on. But the italicised tenses give us twenty-seven third-order temporal attributions. The paradox can, however, be reformulated with respect to these twenty-seven attributions, and the same response is available: no event has them all at once, but has them successively. Spelling out the relevant notion of succession will give us eighty-one attributions, and so on. The passage of time thus seems to give rise to an *infinite regress*: an infinite series of temporal attributions at higher and higher orders.

Now, infinite regression is not always a problem. There are some infinite series that are completely well-behaved and give no reason for concern (think of the manner in which the successor function in number theory (i.e. X comes after Y) gives rise to an infinite sequence of numbers (e.g. the natural or 'counting' numbers 1, 2, 3, 4 ...). The infinitely expanding series of temporal attributions is troubling because *at each level* of the series a paradox can be formulated. That is, when there are three temporal attributions, a paradox can be formulated for those three attributions. Avoiding the paradox requires rising up to the next level. But a paradox can be formulated there, and so on forever. There is a sense in which the paradox is never completely solved: it keeps revenging against us as we move up through this infinitely expanding sequence of temporal attributions.

Actually, it pays to be a bit careful here. Let us differentiate between a vicious regress and a non-vicious regress. Roughly, a regress is vicious with respect to a problem P if the problem at level n can only be solved at level n+1. A regress is non-vicious with respect to a problem P if the problem at level n can be solved at level n, but the solution forces the existence of another level, n+1, and the problem can be reformed at n+1. Intuitively, the concept of a vicious regress just articulated is tracking any problem that is only really solved in the infinite limit, if at all. The concept of a non-vicious regress is any problem that gives rise to a regress in the solution of that problem, but where the solution itself is not obtained only within the infinite limit but, rather, is obtained at any finite step in the regress.

With these sharper definitions in hand, is McTaggart's paradox vicious or not? Maybe not. At each level, the paradox seems to get resolved. True, the paradox can then be reformulated at the next level. But the problem itself is still being solved, it is just that the solution forces an infinite series. McTaggart's paradox, in other words, does not seem to be solved only in the limit. It seems to be solved at any finite step in the regress, it is just that the problem can be infinitely reformulated. Notice, however, that it's a slightly different problem each time (the difference being the number of distinct temporal attributions that are used to formulate the problem).

One response to McTaggart's paradox, then, is to deny the viciousness of the regress. Another response is to deny that the paradox hits every dynamic theory of time. As it was formulated above and, indeed, as McTaggart himself formulated it, the paradox focuses on the properties of being present, being past and being future. As we have already discussed, however, not every dynamic theory of time conceptualises temporal passage in these terms. So perhaps the paradox can be avoided by simply shifting to a different dynamic theory.

In the first instance, at least, this suggestion is no good. There is nothing essential about the use of properties in the formulation of McTaggart's paradox. Every dynamic theory of time maintains a distinction between past, present and future. While the nature of this distinction changes, every dynamic theorist wants to be able to say things like 'dinosaurs are past' and 'the big crunch is in the future'. But these statements are enough to get the paradox going. For every dynamic theorist will accept that 'dinosaurs are both past and present' is incoherent. And every dynamic theorist accepts that things shift from being future, to being present, to being past. So every dynamic theory faces the problem.

Still, one might disagree, at least to a certain extent. While the paradox strikes most dynamic theories of time, one might argue, it leaves presentism untouched. Why so? Well, in order for the paradox to succeed, we need to be able to say, truly, that there are some things that are both past and present (or some other incoherent temporal attribution). But there are no past things, if presentism is true. So nothing is ever both past and present. The same broad response applies to any set of incompatible temporal attributions: only present entities exist. So there is never going to be a case of a thing that resides under multiple temporal attributions, let alone incompatible ones.

A version of the paradox can, however, be reformulated against presentism as well. Here's the basic idea. For the presentist, only present entities exist. Now, imagine the entire universe as a photograph at some present moment. The photograph would show present entities existing at a particular time, t_1, and would show nothing existing at any other time. It would be a photograph in which pretty much everything is black, except for the thin sliver of the present. But, the presentist maintains, the present changes as time passes. To represent the passage of time, we can imagine a series of photographs. Each photograph represents the present moment.

The first photograph represents the present moment at t_1. The next represents the present moment at t_2 and so on until the end of the universe.

All of these photographs represent the same universe. Accordingly, the presentist is ultimately going to have to combine all of these photographs into a single image, one that represents reality. But how do we do that? Suppose that we have just two photographs. In the first photograph t_1 exists and is present, but t_2 is future and is not present. In the second photograph, t_2 exists and is present, but t_1 is past and is not present. When we try to combine these photographs into a single consistent image, we end up with contradictions: t_1 both exists and does not exist; t_2 both exists and does not exist. The only way to avoid this outcome, it would seem, is to say that both t_1 and t_2 exist, or that both fail to exist, and reconcile the two photographs that way. But to take the first option is to admit that more than a single present moment exists, and to take the second is to admit that nothing exists. Neither option is a way of recovering presentism.

A similar argument can be used against any dynamic theory of time. Each dynamic theory will treat the world as a series of photographs; in each photograph the world looks different. Reconciling the series of photographs into a single, consistent image of the world will force the types of contradictions that seem to arise from McTaggart's paradox. What's nice about this way of stating the paradox is that it doesn't rely on any regresses. The point is simply that the dynamic theory of time does not seem to admit of a single, consistent account of the universe. So the worries about whether McTaggart's paradox involves a regress of some kind can reasonably be set aside. Of course, the dynamic theorist might, at this point, simply accept this consequence. Several philosophers have recently considered (or defended) views according to which there simply is no single, consistent, model of our universe: instead there are many inconsistent 'fragments' or 'perspectives' on the universe. Each fragment is consistent, but it is not possible to put these fragments together into a single model or picture of our world. If the dynamic theorist is willing to accept some sort of fragmentalism, or radical perspectivalism, then she has another way of resisting McTaggart's paradox. This is not, however, an avenue we will pursue here.

2.4. Time and Hypertime

So far we have looked at some different accounts of what temporal passage is, and have considered two of the chief difficulties facing any account of temporal passage: the rate of flow argument and McTaggart's paradox. In this section we will look at a unified solution to both problems. Consider, again, the rate of flow argument. The core problem here related to the fact that temporal passage cannot be reasonably understood as a kind of change. Change, it would seem, requires a variation of one quantity against another. But there is no second dimension of variation against which we can measure the type of change that is posited by the dynamic theory of time.

But what if there were some extra dimension of variation? Suppose that we have one time dimension, time, and a second dimension, hypertime. If we have a second temporal dimension, then it seems possible to define the rate of time's passing in a meaningful manner. What we might say is that the present in time, t, moves at one second of time per one second of hypertime, ht. In other words, we may be able to use a second temporal dimension to provide a meaningful rate of passage for time, in much the same manner that we might give an analysis of velocity in terms of change in spatial location over time.

Positing hypertime also provides a solution to McTaggart's paradox. The most straightforward way to see this is to focus on the generalised form of the paradox discussed at the end of the previous section. Consider, again, a presentist universe. As before, let us imagine such a universe as a sequence of snapshots. Let us suppose, for simplicity, that there are just four snapshots: S_1, S_2, S_3 and S_4. In S_1, t_1 is present and t_2, t_3 and t_4 are future. So only t_1 exists. In S_2, t_2 is present and t_1 is past while t_3 and t_4 are future. In S_3, t_3 is present, t_1 and t_2 are past and t_4 is future. In S_4, t_4 is present and t_1, t_2 and t_3 are past. In each snapshot, only a single time exists. If we try to compile the snapshots into a single, flattened image, we get contradictions: t_1 both exists and does not exist; t_2 is both present and past, and so on. But now suppose that instead of combining the snapshots into a single, flattened image, we stack them, one atop the other. We then treat the entire stack as our 'picture' of the universe.

This is, effectively, what the addition of hypertime allows us to do. In order to combine photographs into a stack, we need to arrange them through a third spatial dimension. In order to arrange our snapshots of the universe into a stack, we need an extra temporal dimension in which to arrange them. With the addition of such a dimension, however, the stack can be produced. So, for example, S_1 corresponds to our first hypertime, ht_1. S_2 corresponds to ht_2. S_3 corresponds to ht_3 and S_4 corresponds to ht_4. If we look along the hypertime dimension, and scan across ht_1 to ht_4, we will see a universe animated. In one moment, the present is at t_1. In another moment, it is at t_2 and so on all the way through to t_4. The universe pictured in such a way is consistent. Nothing is ever both present and past, at least not at the same hypertime. For instance, at ht_1, t_1 is present. At ht_2, t_2 is present. But at ht_2 it is not the case that t_1 is also present. Rather, at ht_2, t_1 is past. And so on for each of the snapshots. The same broad solution will work for any dynamic theory of time. For any dynamic theory of time the sequence of snapshots can be consistently arranged along a hyper-temporal dimension so as to avoid attributing incompatible temporal aspects to any one thing.

The appeal to hypertime, then, has a lot to recommend it. Unfortunately, it faces some fairly serious problems. First of all, there is very little evidence that our world has two temporal dimensions. Indeed, our best current physics is not obviously compatible with such an idea. General relativity uses a four-dimensional geometry to describe the universe. In this four-dimensional geometry, there is only a single dimension of time. Adding an extra dimension of time would force us to radically redevelop the geometry. Since gravity is, effectively, a function of the topology of a

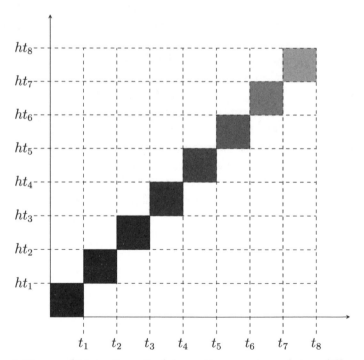

Figure 5 Time and Hypertime: Each hypertime corresponds to a different snapshot, enabling all of the snapshots to be reconciled into a single consistent picture of reality.

four-dimensional universe, altering the dimensionality of our universe may have profound implications for the nature of gravity. At the very least, the differential equations that correspond to general relativity would need to be rewritten, with an added dimension of freedom. Until this is done it is quite unclear how we might add a second temporal dimension into our understanding of the universe.

Second, there are substantial questions to be asked and answered surrounding the nature of hypertime itself. Is hypertime static or dynamic? Suppose that it is dynamic. If it is dynamic, then we will require a third dimension of time in order to avoid two-dimensional versions of the rate of flow argument or McTaggart's paradox. If three-dimensional time is also dynamic, then we will require a fourth dimension of time and so on *ad infinitum*. This regress is vicious: the rate of flow argument and McTaggart's paradox are only resolved in the infinite limit.

Suppose, then, that hypertime is static. If hypertime is static, then the rate of flow argument and McTaggart's paradox are put to rest. But then one wonders what motivation one might have for adopting the dynamic theory of time. After all, the metaphysical foundations of the static theory have been ushered into the dynamic theory as well. Both theories ultimately stand on a static conception of the universe. But if both theories ultimately stand on a static conception of the universe, then why not just simply say that time is static and be done with it? Indeed, simplicity would seem to demand that we do away with the dynamic aspect of time and this extra temporal dimension and just go with what we know: a single, static dimension. A dynamic theorist will no doubt respond that the dynamic aspect of time is needed to do justice to our experiences of time. The static aspect of time alone is not enough to do justice to experience. Once again, then, we see the prospects for the dynamic theory of time come to hang on the weight of an argument that moves from the experience of time to the metaphysics of time. It is to this issue that we will turn, in Chapter 3.

2.5. Summary

In this chapter we have provided an overview of the two main problems for the idea that the passage of time is an objective, mind-independent phenomenon. Along the way, we have gained a better sense of what it might mean to say that time passes, as well as the way in which the passage of time is conceptualised differently on different versions of the dynamic theory. The core points covered in this chapter are as follows:

(1) The passage of time can be understood in terms of a moving present.

(2) The moving present can be understood in terms of a shift in the monadic properties of presentness, pastness and futurity or in terms of a changing ontology.

(3) The rate of flow argument challenges the idea that time passes by showing that there is no meaningful rate at which time passes.

(4) McTaggart's paradox challenges the idea that time passes by maintaining that the passage of time leads to some form of inconsistency, e.g. inconsistent temporal attributions for events.

(5) The dynamic theorist may need to take the flow of time as a primitive notion in order to avoid these difficulties.

(6) The dynamic theorist might appeal to hypertime to avoid the problems, but then she must explain how hypertime works, and must reconcile it with our best current physics.

2.6. Exercises

i. Try to come up with a rate of change that we can apply to the passage of time that is meaningful.

ii. Consider a view according to which future and present entities exist and past entities do not. What notion of temporal passage would we need for such a view?

iii. Try to describe the difference between a vicious and a non-vicious regress. Provide an example unrelated to time of a regress of each kind.

iv. Consider the idea that temporal passage is a primitive notion. What do you think this means? Do you think it is plausible to think of the passage of time in this way?

v. Identify two difficulties with the idea that the passage of time can be reduced to the direction of time.

2.7. Glossary of Terms

Hypertime
A second temporal dimension.

Infinite Regress
An infinite series generated by repeatedly applying some condition, or repeatedly applying a problem.

Non-Vicious Regress
A regress is non-vicious with respect to a problem P if the problem at level n can be solved at level n, but the solution forces the existence of another level, n+1, and the problem can be reformed at n+1.

Ontological Change
A change in what exists and/or the properties that existing things possess.

Temporal Passage
The flow of time, often understood in terms of a moving present.

Vicious Regress
A regress is vicious with respect to a problem P if the problem at level n can only be solved at level n+1.

2.8. Further Readings

H. Dyke (2011) 'Time, Metaphysic of', in *Routledge Encylopedia of Philosophy*, https://www.rep.routledge.com/articles/thematic/time-metaphysics-of/v-2. Section 6 of Dyke's Routledge entry provides a fairly accessible introduction to McTaggart's paradox.

S. Prosser (2016) 'The Passage of Time' in Heather Dyke and Adrian Bardon, eds, *A Companion to the Philosophy of Time* (Wiley-Blackwell), pp. 315–27. A useful introductory reading of the idea that time passes.

J. J. C. Smart (1949) 'The River of Time' *Mind* 58 (232): 483–94. This paper was the first to really articulate the rate of flow argument. Although it is not introductory it is fairly accessible.

J. M. E. McTaggart (1908) 'The Unreality of Time' *Mind* 17 (68): 457–74. This paper is the first to outline a paradox for the A-series. It is also the origin of the A-series/B-series distinction. The paper is not introductory and is really for more advanced students.

D. Williams (1951) 'The Myth of Passage', *The Journal of Philosophy* 48 (15): 457–72. Also for the advanced student, this paper presents a beautifully written and lucid argument against the objective reality of temporal passage.

3

The Experience of Time

In the previous chapter we considered temporal passage. We looked at two central arguments against the idea that the passage of time is an objective, mind-independent feature of reality. In this chapter we will look at the case in favour of the idea that the passage of time is real. The central case in favour of the reality of temporal passage is based on experience. As such, in this chapter we will look in detail at the experience of the passage of time and critically assess the extent to which experience can be used as a guide to the nature of time.

3.1. Temporal Phenomenology

Each of has conscious experiences. We see red; we feel the wind in our hair; we taste vegemite; we smell roses. There is something that it is like to have such experiences. Let's call mental states *phenomenal states* if those mental states are such that there is something that it is like to have them. Further, let's call the *what it is like* to have those phenomenal states their *phenomenal character*. Different phenomenal states have different phenomenal characters. It is different to experience pain than it is to experience pleasure; it is different to experience redness than blueness, and so on. Many philosophers think that for some phenomenal states, not only is there something it is like to be in those states, but also that to be in those states is to be in a state that is *as if* something is the case. For instance, if Sara has a red ball experience it is *as if* there is a red ball to her. Her experience has a certain kind of representational content. It represents that the world seems to be as though it contains a red ball. Usually this is because it does, in fact, contain a red ball. Sometimes it is not; if Sara is hallucinating the presence of a red ball, or she is seeing a blue ball in coloured light, then although her experiences are as if there is a red ball, there is no red ball. Let's call this '*seeming as though the world is as if...*' the *phenomenal content* of a phenomenal state. The phenomenal content of Sara's red ball experience is *as if* there is a red ball. It seems to Sara as though there is, *in the world,* a red ball; the phenomenal state *represents* there being a red ball.

Each of us has temporal phenomenal states and, plausibly, some of these states have a certain phenomenal content. It seems to us as though the world has a certain sort of temporal structure. It seems to us as though events occur in a certain temporal order, for a certain duration, and as though they are separated by a certain duration. It seems to us as though some events occur at the same time, and others occur before, or after, other events. According to some philosophers it also seems to us as though time passes or flows. It seems as though the future is coming ever closer until it becomes the present, which is immediate and especially real. It seems as though the present then recedes into the past where it is fixed and unreachable, and thereafter recedes ever further into history. It seems as though time flows over us, or that we move through time. Let's call however the world seems to us, in these 'timey' ways, our *temporal phenomenology.*

There is scope for disagreement about exactly what our temporal phenomenology is like, that is, about what phenomenal content it has. As we will see, not everyone thinks that our temporal phenomenology is such that it seems as though time passes. It is this purported aspect of our temporal phenomenology with which we will principally be interested in this chapter. That's because, first, there is widespread agreement that it seems to us as though events happen in an order, and with a certain duration, so there is little scope for philosophical dispute in this regard. Second, there are, potentially, metaphysical implications that follow from the dispute regarding whether or not our temporal phenomenology is as of time flowing or passing. It is these metaphysical implications we consider next.

3.2. The Argument from Temporal Phenomenology

Suppose it's true that it seems to us that time passes. That is, suppose it's true that our temporal phenomenology has the phenomenal content *as of* time passing. Then there is an argument from us having this temporal phenomenology, to the conclusion that time really does pass.

> **Argument from Temporal Phenomenology**
> [P1] We have experiences as of the passage of time.
> [P2] If we have experiences as of the passage of time, then any reasonable explanation for this relies on the passage of time being an objective feature of reality.

Therefore,
[C] The passage of time is an objective feature of reality.

Before we consider this argument in more detail, it is worth considering what the argument would show, were it sound. Plausibly, it would show that, actually, there is temporal passage. That is, it would show that some version of the A-theory is actually true, and so any adequate extensional theory of time must be A-theoretic. It would not, however, seem to show that we should endorse a constitutive A-theory of time. For it would not show that temporal passage is essential to time itself. [C] could be true, and yet there be worlds with a very different phenomenology in which time seems to be static, and in which time's seeming that way is explained by time's being static.

The inference from [P1] and [P2] to [C] involves what is known as inference to the best explanation. The idea, very roughly, is that we ought to accept the best explanation of some phenomenon, and hence we ought to accept whatever ontology the best explanation requires. So, for instance, if the best explanation for the presence of rustling in the walls, and missing cheese, and gnawed belongings, is that there are mice in the walls, then this is the explanation we should accept. Moreover, having accepted that explanation we need to posit mice: that is, we need to accept that mice exist. Likewise, if the best explanation for the ills of the world is that a nasty demon is going around causing earthquakes, and famine, and unemployment, then we should accept that explanation, and in so doing, accept that such a demon exists.

The argument from temporal phenomenology presses us to accept that there is temporal passage on the basis of its alleged explanatory indispensability. The thought is that we must provide some reasonable explanation for our experiences as of passage, and that any such explanation will inevitably rely on the existence of temporal passage. [P1] is typically supported by a direct appeal to phenomenology: surely you can *just tell* that your temporal phenomenology represents the world as containing passage. [P2] is typically supported by two thoughts: that the presence of passage would render our temporal phenomenology veridical, thus providing a simple and reasonable explanation of it; and that no purported explanation of our temporal phenomenology that *doesn't* rely on passage could be a reasonable one. If the existence of temporal passage provides the *only* reasonable explanation for our experiences as of temporal passage, then *a fortiori* it provides the *best* explanation. If it does, indeed, provide the best explanation then we ought to conclude that there is temporal passage. So if [P1] and [P2] are true, then we should conclude that [C] is true. We have already seen, in Chapter 2, that problems arise for views that accept [C], and we will consider a further reason to reject [C], coming from physics, in Chapter 4. But the argument from temporal phenomenology gives us reason to accept [C]. In this chapter we take the perspective of someone who is unsure whether to accept [C] on the basis of the argument from temporal phenomenology, by looking to see whether there are plausible ways to resist [P1] or [P2]. At the end of section 3.4 we consider the extent to which any of these undermine the argument from temporal phenomenology.

3.3. Rejecting [P1]

We will begin by considering views that reject [P1]. Such views fall into two broad categories: cognitive error theories and no content theories. We will consider each of these in turn.

3.3.1. Cognitive Error Theory

Cognitive error theorists reject [P1]: they hold that we do not have experiences as of time passing. That is, they reject the claim that we have any experiences whose phenomenal content is as of time passing. Nevertheless, cognitive error theorists maintain that our temporal phenomenology is *veridical*. A representational state is veridical just in case what it represents is the case. So if it seems to Sara that the world is as if there is a chair and table opposite her, and if there is indeed a chair and table opposite her, then Sara's experience is veridical. By contrast, if there is no chair and table opposite Sara, then her experience is *non-veridical*, or *hallucinatory*. If our temporal phenomenology in general is veridical then it represents that the world is some way, and the world is that way.

For instance, perhaps in addition to our temporal phenomenology representing that events occur in a particular order, with a particular duration, it also represents that some of those events are contemporaneous both with one another and with the experience, as well as representing that these events are constantly changing. We will consider some alternative views about what content our temporal phenomenology has very shortly. What matters for now is just that this minimal description of our temporal phenomenology appears to be mostly veridical (of course, we can and do suffer temporal illusions, in which things appear to be a way that they are not: the point here is just that for the most part this is not happening).

According to cognitive error theory, despite being veridical, our temporal phenomenology is not as of passage. While our temporal phenomenology does have phenomenal content – it does represent that things are thus and so – it doesn't represent that time passes. In fact, according to cognitive error theorists, this is precisely why our temporal phenomenology is veridical: it is veridical because there is no temporal passage, and our temporal phenomenology does not have phenomenal content as of there being passage. If it did have that content, then, since there is no passage, it would fail to be veridical.

Of course, it's not enough to simply say that our temporal phenomenology is not as of time passing. After all, there seems to be widespread belief that it *does* seem as though time passes. One couldn't simply stipulate that it doesn't seem as though there are chairs, given that so many people seem to think it does seem that way. So what can the cognitive error theorist say? She will concede that there is a widespread *belief* that it seems as though time passes. But, she will say, this belief is false. We are subject to a *cognitive error*: a false belief. So although [P1] is false, we falsely believe it to be true, because we falsely believe that we have such experiences.

How on earth could we come to have a false belief about what our own experiences are like? Well as we will see, that's not so strange after all. There is lots of empirical evidence of just these sorts of false beliefs. How could it happen in this particular case? There are roughly two explanatory strategies the cognitive error theorist might employ. We call these the misdescriptionist and the inferentialist strategies respectively. According to the misdescriptionist we *misdescribe* our temporal phenomenology as being as of time passing, and it is this misdescription that causes us to believe that our phenomenology is as of temporal passage. According to the inferentialist we falsely believe that our phenomenology is as of time passing because some (probably sub-personal) mechanism in our brain infers *from* our belief that our temporal phenomenology is a cause of our belief that time passes *to* the belief that our temporal phenomenology is as of time passing. We will consider each of these strategies in more detail below.

Let's consider the misdescriptionist strategy first. Why might we misdescribe our own phenomenology, and even if we did, why might this result in us forming false beliefs about the content of that phenomenology? Here's one hypothesis. Certain natural language constructions and expressions predispose us to (i.e. make it more probable that we will) conceptualise the world in certain ways. There are plenty of examples of what we might call passage-friendly language in English. For example, people often use phrases such as 'time is passing'; 'Wednesday is coming up'; 'once we get past next week', 'time flows like a river' and so on. When we speak, hear or read such things, we often reason on the basis of them. One might say that an implicit dynamic theory of time is embedded in our natural language.

We know, from psychological research, that using or hearing certain phrases involves entertaining, at least briefly, the ideas expressed in those phrases. So if our natural language is full of phrases about how time passes, we should expect that using or hearing those phrases will involve entertaining the idea that time passes. There is also plenty of evidence that using language in these ways can result in individuals not only having certain concepts, in this case the concept of temporal passage, but that in deploying those concepts individuals come to understand and navigate the world by using those concepts.

For instance, there is evidence that the way in which linguistic communities write language affects the way that they conceptualise time. Mandarin speakers (who write vertically) are much more likely than English speakers to use a vertical axis when mapping out time, and while both English and Mandarin speakers appear to represent time as running along a left-to-right axis, Mandarin speakers also show evidence of a vertical arrangement of time, with earlier events represented higher. English speakers show no such pattern. Moreover, those who are bilingual in both Mandarin and English show a bias in favour of the language in which they are most proficient. That is, those who are most proficient in Mandarin are more likely to arrange time vertically, and those who are more proficient in English are more likely to arrange time horizontally. When tested in English, bilingual individuals are more likely to arrange time horizontally, and more likely to arrange it vertically when tested in Mandarin. Moreover, Mandarin speakers possess far more vertical metaphors than English speakers, suggesting that there is relation between the way in which a language is represented in written form, the metaphors speakers of the language deploy, and the way in which speakers represent the temporal dimension.

So the misdescriptionist might hold that conceptualising the world in ways that are more consistent with an implicit dynamic theory will cause us to *describe* our temporal phenomenology using the passage-friendly expressions found in natural language. Why should this cause us to believe that our phenomenology is as of time passing?

Well, notice that most people describe a sunrise as just that: the rising of the sun. No doubt if pressed they would say that it seems to them as though the earth stands still, and the sun rises. Although most of us know that in fact it is the earth that is rotating and not the sun that is moving, this knowledge hasn't shown up in our natural language expressions. We still talk about sunrise and sunset, and about the sun rising in the east, and setting in the west. We talk of the sun 'going down' in the evening and so on. The implicit theory that the earth stands still, and the sun moves, is embedded in natural language. Plausibly, then, it is this embedding that leads us to *describe* the experiences we have as being ones in which the sun rises. After all, we could surely have the very same experiences and describe them as experiences as of the earth rotating and the sun standing still. Things wouldn't *seem* any different, but we would describe the phenomenology as being *as if* the earth rotates. The misdescriptionist thinks this is exactly what is happening in the case of our temporal phenomenology. She thinks that the world seems a certain way to us – it seems as though events happen in an order, and that at each moment we remember more events than at previous moments, and it seems as though decisions that were previously open become closed as we choose to do one thing or another and then do that thing – but because our language is filled with passage-friendly expressions we *describe* that phenomenology using expressions such as 'my birthday is nearing' which make us come to believe that it seems to us as though time passes.

Of course, for the misdescription strategy to be plausible it needs to be that whatever the content of our temporal phenomenology, it *could* mistakenly be described as having content as of passage. It is, however, noteworthy that descriptions of our temporal phenomenology are often quite vague, and involve a lot of metaphorical language. The cognitive error theorist will point out that if the content of our temporal phenomenology is sufficiently amorphous then it is plausible that when our natural language is imbued with passage-friendly expressions that make it easy to conceptualise the world in terms of passage, we will describe our temporal phenomenology in those terms. Moreover, the cognitive error theorist can point out that we know that the way people describe the content of a phenomenological state is highly dependent on their beliefs about the cause of that state. For instance, it's common to interpret experiments on emotion as showing that subjects can be brought to describe a phenomenological state with essentially the same phenomenal character as being anger, love or fear, depending on the context in which the state is elicited and depending on what the subjects believe to be the cause of that state. On this way of interpreting the evidence, different beliefs about the cause of some phenomenal state bring it about that the phenomenal state is described as being, say, love, rather than fear. At the very least, then, there is evidence that different background beliefs can affect the way we describe the self-same phenomenology, and, indeed, that the very same phenomenology can be described *very* differently.

So, what might our phenomenology be like, such that we misdescribe it as being as of passage? Recently, some philosophers have hypothesised that conscious perception is determined by whatever hypothesis about the cause of sensory input has the highest posterior probability – what they call the 'winning' perceptual hypothesis. The thought is that one's brain generates a number of hypotheses about what it is one might be seeing, based on previous evidence. Whichever of these hypotheses has the highest probability given the sensory data one receives, that hypothesis is the one that wins. That is what the brain takes a person to be perceiving and thus what it seems to that person that they are seeing.

Winning perceptual hypotheses have only a limited lifespan; a perceptual hypothesis only counts as the winner for a short period of time. This is because as soon as one's perceptual system settles on a winning hypothesis it begins decreasing the probability of that hypothesis, *distrusting it*, as the best account of one's current sensory input. The perceptual system does this because the external world changes, and the best perceptual hypothesis to explain the incoming sensory stimuli at one time is less likely to be the best hypothesis about incoming sensory stimuli at later times.

As a result of our perceptual system *distrusting the present hypothesis* it begins to seek out alternative perceptual hypotheses to explain incoming sensory input. These alternative perceptual hypotheses are hypotheses about some possible state of affairs that may or may not be occurring *at the very time of the hypothesis*. Our perceptual system is continually placing the winning perceptual hypothesis head-to-head with alternative perceptual hypotheses about what is happening. It has been suggested that this process is responsible for certain aspects of our temporal phenomenology. In particular, some have suggested that this mechanism is responsible for us having temporal phenomenology as of passage. But really, there is nothing about the perceptual process just described that is inconsistent with the cognitive error theory. The cognitive error theorist can insist that our phenomenology is really as of constant change and evolution; it is really as of the world meeting, or failing to meet, our changing expectations of what is around us. This phenomenology could easily be misdescribed as being as of time passing given our use of passage-friendly language. Then it will be very easy to come to believe that it seems to us as though time passes.

That brings us to the second strategy the cognitive error theorist might adopt for explaining why we mistakenly believe that it seems to us as though time passes: the inferentialist strategy. The inferentialist aims to explain why we have this false belief about our phenomenology by suggesting that we infer that this is the content of our phenomenology. We make this inference because we already believe that our temporal phenomenology causes our belief that time passes. The first thing to note about this strategy is that it appeals to us having a belief that time passes. If the strategy is to succeed, the inferentialist needs to explain why we believe that time passes, despite the fact that, according to her, it does not. Moreover, she needs to explain why we have that belief without suggesting that we have the belief because it seems to us that time passes.

In effect, the inferentialist wants to reverse the order of explanation here. Rather than holding that we believe that time passes because it seems to us that time passes, she wants to suggest that because we believe that time passes, we infer that it seems as though time passes. That means she needs to explain why we believe that time passes, without appealing either to time's passing, or to its seeming that time passes. How might she do this? There are various options, and we have already met one of these in our discussion of the misdescriptionist strategy. Recall that the misdescriptionist claims that it is because our language is imbued with passage-friendly expressions that we misdescribe our phenomenology as being as of passage. The inferentialist can offer something like the same account to explain why we come to believe that time passes. She can suggest that we form that belief because an implicit dynamic theory of time is embedded in natural language expressions.

Whatever story the inferentialist offers about how we come to believe that time passes, she will need to appeal to our having that belief to explain why we come to falsely believe that *it seems to us* that time passes. The idea, very roughly, is that we have a general background belief that perceptual states cause certain kinds of beliefs. For instance, Sara believes that there is a chair opposite her, and she is having a perceptual state as of there being a chair opposite her. It is natural to suppose that the *reason* she has the belief that the chair is opposite her is that she is having a perceptual state as of there being a chair. Belief states caused by perceptual states usually have the content they do – are about the things they are about – because they are caused by perceptual states which are, in turn, caused by, in this case, chairs. Indeed, if our perceptual states and belief states didn't work like this, then it would be impossible to navigate the world around us: Sara's having a perceptual state as of there being a tiger in front of her wouldn't cause her to believe there to be a tiger in front of her which would, in turn, likely cause her to be eaten by that tiger.

Furthermore, according to most philosophical accounts, mental states get to have the content they do – get to be about the things they are about – due to the ways in which those mental states are connected to the world. While different theories of content differ over exactly what is required in order for a mental state to be about, say, a cow, they almost all agree that some important component of the story will involve causal connections between the mental state and cows. The reason Sara's perceptual state is about a chair, and not an elephant, is (at least in part) because it is typically caused by chairs, and not by elephants. Likewise, the reason her belief that there is a chair in front of her is a belief about a chair and not an elephant is (at least in part) because the belief is typically caused by chairs, not elephants.

The inferentialist will marshal these general features of our cognitive system alongside these general features of the ways in which our mental states get their content. She will suggest that we each come to believe that our belief that time passes is caused by our temporal phenomenology, just as Sara's belief that there is a chair opposite is caused by her chair phenomenology. The inferentialist can then suggest that, because we believe that our temporal phenomenology causes our belief that time passes, we infer (probably sub-personally rather than consciously) that our temporal phenomenology is as of time passing. For in general that's a good inference: if Sara's phenomenology causes her belief that there is a chair opposite her, then it's a fairly safe bet that her phenomenology is as of there being a chair opposite her. With respect to the passage of time, the analogous inference is a bad inference because time does not pass, and it causes us to come to falsely believe that our temporal phenomenology has a content that it does not.

That brings us to the end of our discussion of cognitive error theory. Recall that cognitive error theorists deny [P1] of the argument from temporal phenomenology. These theorists hold that our temporal phenomenology has phenomenal content – it does indeed seem to us as though the world is some way – it is just that it does not seem to us as though time passes. Other theorists, too, deny [P1], but in doing so they reject the contention that our temporal phenomenology has any content at all: they deny that our temporal phenomenology represents to us that the world is a certain way. We will consider this view in the next section.

3.3.2. No Content Theory

To understand the no content theory we need to return to our earlier discussion of phenomenal character and phenomenal content in section 3.1. We noted there that it is widely accepted that at least some of our phenomenal states have phenomenal content: that they represent that the world is some way or other. One might, however, deny this. Instead, one might think that purely phenomenal states do not represent the world at all. To be sure, there is something that it is like to be in those states. And, to be sure, we have representational states: mental states that represent that the world is some way or other. It presently seems to Sara as though there is a chair opposite her, and that seeming is, indeed, representational. But the seeming is not some pure phenomenal state. The pure phenomenal state, if there is one, presents her with nothing but shapes and colours. It seems to her that there is a *chair* opposite because she has a whole lot of complicated hypotheses about what is opposite. Or, at least, her sub-personal perceptual system has those hypotheses. The reason it seems that there is a chair opposite has a lot to do with her having a whole bunch of representational states: background beliefs, perceptual hypotheses, and so on.

The no content theorist thinks that there is a way things seem to Sara, temporally speaking. But the way things seem to her is the result of a complex interaction between her phenomenal states and her representational states. There is no purely phenomenal state that represents that the world is as of time passing. When we say that a phenomenal state is one in which it seems as though there is a chair, or as though time passes, we are reporting on the content of a representational state which is the product of our phenomenal state plus some other features of our cognitive system. When Sara says that it seems to her as though there is a chair opposite, she is not reporting the content of a phenomenal state. Instead, its seeming that there is a chair opposite is the product of having certain phenomenal states, and her conceptualising or understanding those phenomenal states through the lens of having certain representation states or hypotheses, to wit, that there is a chair opposite.

In the case of temporal phenomenology the idea is that we each have both phenomenal states and representational states, where these latter might include an explicit belief that time passes, or might include other beliefs or representations or conceptualisations of the world. It is these additional representational states which, in conjunction with our phenomenal state, produce in us a representational state as of time passing.

Notice that this view looks rather like the misdescriptionist view. The key difference is that for the misdescriptionist our phenomenal states do have content. They do represent that the world is thus and so. The misdescriptionist's contention is that the presence of certain factors causes us to misdescribe what that content is. By contrast, the no content theorist maintains that our phenomenal states do not, in themselves, have content at all, though of course they do have phenomenal character. There is something it is like to be in those states. But those states don't represent to us that the world is some way until they are 'overlaid' with additional representational states. If we have false beliefs that time passes, or if our language embeds passage-friendly phrases, then these factors might result in us coming to think that our temporal phenomenal states represent that time passes when in fact they *represent* nothing at all.

That brings us to the end of our discussion of views that reject [P1] of the argument from temporal phenomenology. In what follows we consider views according to which [P1] is true: it does seem to us as though time passes. Because these views concede [P1], they each seek to avoid the argument from temporal phenomenology in some other fashion.

3.4. Rejecting [P2]

In this section we consider a view that rejects [P2]. Phenomenal illusionists accept [P1] but reject [P2]. According to the phenomenal illusion view, it does seem to each of us as though time flows, but this is an illusion, since time does not flow. Yet, they maintain, we have a perfectly good explanation for the presence of this phenomenal illusion, so we have no reason to posit temporal passage.

Before we discuss the phenomenal illusion view it is worth briefly returning to our earlier discussion of cognitive error theory in order to clarify the connection between that view and the phenomenal illusion view. 'Cognitive' is sometimes used (particularly in psychology) when appended to 'mechanism' or 'process' to pick out some set of mechanisms or processes that, very loosely, involve thought, or representation, or the manipulation of representations. Given this use of the term, it would be easy to think that cognitive error theorists hold a dual thesis: that we are subject to a false belief about the content of our temporal phenomenology, and that the mechanism responsible for that false belief is cognitive. By parity, then, phenomenal illusionists would be those who hold the dual thesis that we are subject to a phenomenal illusion, and that the mechanism responsible for our phenomenal error is non-cognitive. But this is a mistake. What distinguishes the phenomenal illusion theory from the cognitive error theory is what that error *consists in*, not what the error is *due to*. It is consistent with the phenomenal illusion theory that the mechanism *responsible* for our having phenomenology as of time passing might be cognitive, in the sense that the mechanism might be loosely involved in thought, representation, or the manipulation thereof. What distinguishes the phenomenal illusionist from the cognitive error theorist is not whether or not the mechanism *responsible* for the error is cognitive or not, but whether the error is an error *in our phenomenology* or an error *in our beliefs*.

The phenomenal illusionist believes there is good evidence that we sometimes experience phenomenal illusions. When we look at the two lines in a Müller-Lyre case, it seems to us as though one line is longer than the other, even though they are the same length (see Figure 6).

Indeed, one line continues to look longer than the other even after we measure them and know them to be the same length. The phenomenal illusionist thinks that our temporal phenomenology is like this. Some feature of our cognitive architecture tricks us so that it seems to us as though time passes. The phenomenal illusionist might even say that our experience as of time passing is like our experience as of the two lines being a different length in the following sense: the reason why we experience the illusion we do is due to some proper functioning of our cognitive architecture.

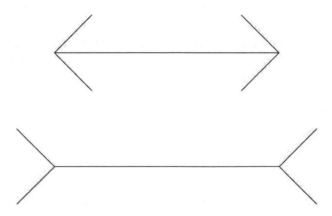

Figure 6 The Müller-Lyre Illusion

It is useful to pause and think about the Müller-Lyre illusion in a bit more detail. There is an explanation for why the two lines seem to be of different lengths. The angles act as depth cues which we associate with three-dimensional scenes. We incorrectly view the image as a three-dimensional drawing, and a size constancy mechanism – a mechanism that allows us to see some objects as being further away, rather than smaller – makes us think that one of the lines is longer because, were the drawing three-dimensional, that line would be further away and hence would, in fact, be longer than the other line. The point is that although our perceptual system gets it wrong on this particular occasion, the error is the result of us having a perceptual system with certain features that, in general, allow it to correctly perceive depth and distance. The phenomenal illusionist might maintain that there are adaptive features of our cognitive architecture that result in it seeming to us as though time passes. There are various different accounts of why this might be. We will consider three such accounts in a moment.

Before we do, however, there is a fairly obvious problem that the phenomenal illusionist must address. Consider the question: what makes it the case that a mental state is a hallucination *of X?* What makes it the case, for instance, that Sara's hallucination is a hallucination of a dancing purple elephant rather than a hallucination of a sad red badger? The intuitive answer is that Sara's hallucination is as of a dancing purple elephant rather than a sad red badger if the content of her representation is as of a dancing purple elephant. That is, the content of her mental state is just what it would be, were there a dancing purple elephant present. What makes the state hallucinatory is just that there is no dancing purple elephant nearby. We noted earlier in this chapter that most theories of representational content hold that causal relations between features of the world and our mental states are at least part of the story about how those mental states come to represent the things they do. Why is Sara's mental state a mental state as of a blue ball? The answer is, at least in part, because it's the type of mental state that is typically caused by the presence of a blue ball.

Prima facie, however, this account of representational content is in tension with the phenomenal illusionist's view. For while the illusionist claims that our experiences represent the world as containing temporal passage, she must deny that these experiences are causally (or otherwise) connected with temporal passage, since she denies that passage exists. According to the phenomenal illusionist we are *always* being tricked.

Can the illusionist make any sense of how we could represent something that does not exist? Well, consider Sara's hallucination of the dancing purple elephant. To be sure, Sara has never been in causal contact with a dancing purple elephant. Dancing purple elephants don't exist. But Sara has been in causal contact with lots of purple things, lots of dancing things, and quite a few elephants. By putting together those representational elements, she can represent a dancing purple elephant. So an appealing strategy for the phenomenal illusionist is to hold that our phenomenology as of time passing is somehow *made up of* representational elements, where each of these elements represents something that does exist, and has the content it does partly in virtue of being caused by that thing.

To provide a sense of how such a story might go, we will consider two different accounts that the phenomenal illusionist might offer. Then we will return to the question of how these accounts explain how we can represent something – time's passing – that fails to exist.

One very general strategy to which phenomenal illusionists appeal is to link our phenomenology as of time passing with our experiences of motion and change. Illusionists of this stripe – call them motion-change illusionists – will think that there is something plausible about the idea that our experiences of motion and change could contribute to, or jointly cause, our phenomenology as of time passing. They are also moved by noticing that a range of studies in psychology and cognitive science show that we can have experiences *as of* motion simply by perceiving static stimuli – i.e. stimuli that are instantaneous or highly temporally localised – flashed at different locations.

According to some cognitive scientists, motion can be conceived of as 'painted' onto our experience of discrete, static, properties. If that is right then an explanation of our experiences as of motion can be obtained by appealing only to the static properties of objects at locations and the static properties of brain structures at times. Why is this important? Well, one might worry that even if one could explain why we have a phenomenology as of time passing by appealing to our experiences of motion and change, this will do us no good at all if motion and change depend on temporal passage, or if our experiences of motion and change depend on temporal passage, by, for instance, depending on the passing of brain events. Whether or not this is the right way to think about motion or the experience of motion is controversial. What matters, for the success of the phenomenal illusionist's project, is whether a correct account of motion and of our experience of motion requires appeal either to temporal passage, or to the phenomenology of passage. Assuming it does not then the phenomenal illusionist has at least the beginning of an account of how we come to have temporal phenomenology as of time passing even though time does not pass.

Motion-change illusionists point to research that shows that there are motion illusions, for example the motion after-effect and phi motion, in which individuals have experiences as of motion when in fact nothing in the external world is in motion. Indeed, it has been found that individuals can have experiences as of motion without also having corresponding experiences as of position change. Since cognitive science not only shows *that* motion illusions arise, but also *explains how* they arise without appealing to the existence of motion, motion-change illusionists argue that if our temporal phenomenology is somehow the result of our motion phenomenology, or the result of our motion phenomenology plus something else, then we can explain how we have that phenomenology in the absence of temporal passage. Motion-change illusionists also appeal to the role of change in our phenomenology.

When we discussed the cognitive error theory we introduced a recent model of perception involving perceptually salient hypotheses (the idea that some perceptual hypothesis about the world is always 'winning'). We suggested that appealing to that model might enable the cognitive error theorist to explain why it is plausible that we could misdescribe the content of our temporal phenomenology as being as of temporal passage when in fact it is not. That, however, is not the only use to which this model of perception might be put. One might, instead, argue that the model explains (or is part of the explanation for) why we have a phenomenology as of passage (even if there is no passage). Recall that according to the perceptual hypothesis at issue, our temporal phenomenology is caused by the perceptual system predicting that the world is a changing place, which in turn leads us to distrust the present perceptual hypothesis. As a result of distrusting the present perceptual hypothesis, the perceptual system continually pushes itself forward into an alternative perceptual hypothesis that best explains the current incoming sensory input. Perhaps, then, the apparent movement of time is caused by our perceptual system continually moving from one perceptual hypothesis to the next, and the constant cycling of hypotheses about what is occurring right now causes us to feel as if time flows.

At any rate, motion-change illusionists typically think that, in some way or other, our experiences as of motion and/or change constitute, or partially constitute, our experience as of time passing, though they may disagree about exactly how or why this occurs.

The motion-change illusion theory is not the only version of phenomenal illusion theory. The second broad strategy available to the phenomenal illusionist for explaining how it is that we might have experiences as of passage borrows an explanation of a different phenomenon: an explanation for why the world seems to us to have an open future (to be ontologically as the growing block theorist supposes it to be) even though some version of the block universe theory is true. The phenomenal illusionist can adopt this basic explanation and repurpose it for her own ends. Here's the idea. At any time each of us has a temporally embedded point of view, which is a representation of time relativised to a particular moment in a psychological history. The temporally embedded perspective is an individual's representation of its history, its memories of memories, its anticipations, its memories of anticipations, and so on. The past, and particularly past decisions, are represented as decided: our beliefs about our past decisions are formed on the same basis as our beliefs about any other past event: by gaining evidence. By contrast, from a temporally embedded perspective looking into the future, the individual represents future decisions as as-yet undecided. Moreover, an individual's beliefs about her own future decisions are not formed by consulting evidence. Sara doesn't work out whether she will have toast or cereal for breakfast tomorrow by looking to see what she had for breakfast on other days of the week: she works it out by deliberating about what she wants to have for breakfast, and then deciding that she will have cereal. For Sara, the only sure way of arriving at true beliefs about what she will do is to let the decision process run its course, and to *decide* what she will do.

The thought is that we can explain why the future seems to us to be open by combining the insight that from each temporally embedded perspective there is an asymmetry with respect to how we represent, and come to know about, past and future decisions, with the insight that in addition to each temporally embedded perspective there is a temporally evolving view. The temporally evolving view is obtained by stringing together the temporally embedded snapshots in the right order. The temporally evolving view consists in the movement through a series of temporally embedded perspectives, such that at each embedded perspective more of the past has been represented, and some of what were previously open future decisions have become fixed past decisions. The phenomenal illusionist might argue that this apparent 'movement' at which future decisions become present, and then past, relative to different embedded perspectives accounts for why it seems to us that time passes.

Whatever exact story the phenomenal illusionist tells, she will no doubt suggest that our having a phenomenology of temporal passage in the absence of there being any temporal passage is the result of some feature of our cognitive architecture that evolved to represent features of our environment that do exist: change and motion are just two hypotheses about what these features might be. So long as the phenomenal illusionist can make it plausible that our cognitive mechanisms generate representational elements which, when combined, lead us to represent that time passes, she has a compelling account of why, despite [P1] being true, [P2] is false. There is a perfectly good explanation for us having the temporal phenomenology we do, one which does not appeal to the existence of temporal passage.

3.5. What Does This Tell Us?

So far we have outlined a number of ways of resisting [P1] or [P2]. But, one might think, unless one already had reason to resist such premises, none of these arguments give us independent reason to think either premise is false. Sure, our temporal phenomenology *could be* systematically illusory. Any of our phenomenology could be systematically illusory: but in the absence of some positive reason to think it is illusory in this manner, we have no good reason to reject [P2]. Equally, it *could be* that we are misdescribing our temporal phenomenology, just as it could be that we are misdescribing a whole range of other phenomenology. But again, in the absence of a positive reason to think that this is so, we have no reason to reject [P1]. So in a way, nothing we have said undermines the argument from temporal phenomenology unless one already has some independent reason to be sceptical that there is temporal passage. Of course, many philosophers take themselves to have such independent reason. So we can think of the arguments just offered as ways of resisting what might *otherwise* seem like a very potent argument in favour of [C]. But if one does not take oneself to have these independent reasons, then at best the arguments considered so far might make one less sure whether to accept [C] on the basis of [P1] and [P2]. That, however, is assuming that although there *might* be some other explanation for why we have the phenomenology we do, other than there being temporal passage, the presence of temporal passage would be a *better* explanation for our having the phenomenology we do, and so absent some good reason to think it's *not* the explanation (i.e. because there is no temporal passage) the simplest conclusion is that it is. In what follows we consider whether, in fact, the presence of temporal passage would be a good explanation for our temporal phenomenology. For if it would not, the alternative explanations for our phenomenology just considered look to be on much firmer ground.

3.6. Passage Theory

Passage theorists accept [P1] and [P2], and hence accept [C]. They think that we have a phenomenology as of passage, and that this is because time flows. Passage theorists accept a dynamic theory of time. Not everyone who accepts a dynamic theory of time is a passage theorist in the sense just articulated. One can accept a dynamic theory of time and reject the claim that the presence of temporal passage is the *only* reasonable explanation for our having the temporal phenomenology we do. Nevertheless, even temporal dynamists who reject the argument from temporal phenomenology almost certainly think that the fact that time passes is a *good* explanation for our having the temporal phenomenology we do. After all, they think this is, in fact, the explanation! So all temporal dynamists seem committed to some claim about the explanatory connection between our having the temporal phenomenology we do, and the presence of temporal passage. Those who are passage theorists endorse a strong claim, namely that the only reasonable explanation for our temporal phenomenology is the presence of temporal passage, while all temporal dynamists must accept at least the weak claim that the presence of passage is a good explanation for our having the temporal phenomenology we do.

This raises the question, if there were temporal passage, would it provide a good explanation for our having the phenomenology we do? If not, all temporal dynamists would appear to be in trouble.

In the next section we consider three distinct kinds of arguments that aim to show that even if there were temporal passage, its presence would not be a good explanation for our temporal phenomenology. These arguments, if sound, show that [P2] in the argument from temporal phenomenology is false. But they do more than that: they also show that temporal dynamists who are not passage theorists face a difficult problem for their view.

3.7. Temporal Passage and Explanation

The passage theorist maintains that the presence of temporal passage is the only reasonable explanation for our having the temporal phenomenology we do. The temporal dynamist maintains that, at the very least, the presence of temporal passage is a good explanation for our having the temporal phenomenology we do. These claims seem *prima facie* plausible. If there is temporal passage we have a very good explanation for why it seems to us as though time passes. But if we really think about it, is the explanation any good? A number of recent arguments suggest not. We divide these arguments into two broad kinds. The first we call arguments from physics because they appeal to certain premises that are supported by contemporary physics. The second we call arguments from metaphysics, because they appeal to certain metaphysical premises. The arguments from physics and the arguments from metaphysics aim to show that the following thesis, which we call the Difference Making Thesis, is false.

The Difference Making Thesis
The existence of temporal passage makes a difference to our temporal phenomenology.

What is the importance of the Difference Making Thesis? Well, consider the following argument:

The Makes No Difference Argument
[P1*] For temporal phenomenology to provide evidence of temporal passage, it must be that the presence of temporal passage makes a difference to our temporal phenomenology.
[P2*] It is not the case that temporal passage makes a difference to our temporal phenomenology.
Therefore,
[C*] Our temporal phenomenology provides no evidence for the presence of temporal passage.

If the argument is sound then [C*] is true. If [C*] is true, then the only reasonable explanation for our temporal phenomenology cannot be the presence of temporal passage. For if X provides no evidence for Y, then it cannot be that Y explains X. So, far from temporal passage being the only reasonable explanation for our temporal phenomenology, it is no explanation at all. So anyone who defends the argument from temporal phenomenology must reject [C*]. Yet [P1*] seems undeniable. Surely if our temporal phenomenology provides us with evidence that there is temporal passage it must be because temporal passage makes a difference to that phenomenology. It must be that the reason our temporal phenomenology has the phenomenal character it does, is because there is temporal passage. That means the defender of the argument from temporal phenomenology must reject [P2*]. She must maintain that the Difference Making Thesis is true, and hence that [P2*] is false. By the same token, the critic of the argument from temporal phenomenology must defend [P2*] and thus argue against the Difference Making Thesis.

In what follows we will follow the path of the critic and look at arguments from physics and metaphysics, which aim to show that the Difference Making Thesis is false and thus that [P2*] is true. If [P2*] is true, then [P2] of the argument from temporal phenomenology is false, and that argument, in turn, is unsound.

3.7.1. Arguments from Physics

Here's the first argument from physics, against the Difference Making Thesis (and hence in favour of [P2*]).

> **The Physical DM Argument**
> [P1] Temporal passage will not show up in any description of the world offered by the physical sciences.
> [P2] If some phenomenon will not show up in any description of the world offered by the physical sciences then that phenomenon makes no physical difference to how things are. Therefore,
> [P3] Temporal passage makes no physical difference to how things are.
> [P4] Anything that makes no physical difference to how things are makes no difference to our temporal phenomenology.
> Therefore,
> [C] If there is temporal passage it makes no difference to our temporal phenomenology.

[P2] in this argument makes the fairly plausible claim that if some phenomenon will not show up in any description of the world offered by the physical sciences, then that phenomenon does not make any physical difference to how things are. The assumption is especially plausible in light of the fact that, for many philosophers, what it is to be a physical property in the first place is to be the kind of property that can, in principle, be detected by the physical sciences. Note that we should *not* read [P2] as the claim that if something makes a physical difference to the world then it should show up in the *current* description of the world offered by the physical sciences. Rather it should be read as the claim that if something makes a physical difference to the world then it will eventually show up in some present or future description of the world offered by the physical sciences.

Why? Well, that's because it would be far too hasty to conclude from the fact that something doesn't appear in our current best physics that therefore it is physically irrelevant. For if this were true then it would be far too easy to offer arguments that are similar to the Physical DM Argument that establish the physical impotence of any new physical feature we might discover that is at odds with our current physics. For example, suppose that Sara discovers a new sub-atomic particle, a squawk. Suppose, however, that it plays no role in current physics. Well, then, it follows from [P2] that squawks make no physical difference to the world. But that is clearly false.

So we should be thinking in terms of completed physics: the physics of the future, when we are considering the Physical DM Argument. Accordingly, some measure of caution is warranted. Physical science is far from complete, and it is a bold claim indeed to suggest that we can have a good sense of what a complete physical science will look like. So it's a bold claim to suggest that we have good grounds to suppose that, whatever it looks like, it won't mention temporal passage. Still, there are considerations in favour of [P1]. Current best physics does not posit temporal passage. Quite the reverse. In Chapter 4 we more fully discuss the connection between time and contemporary physics. For now, we can just note that contemporary physics is quite hostile to the idea of temporal passage. It is certainly no part of general and special relativity that time passes. Yet these are very well confirmed theories. So it would be surprising if a description of our world couched in the language of the physical sciences made mention of temporal passage, since that would mean that some of our best confirmed theories are quite radically false.

Still, defenders of the argument from temporal phenomenology have some cause for hope. As we also discuss in Chapter 4, quantum mechanics and general relativity are incompatible. So we know that *something* must give. Since some ways of reconciling the two theories posit temporal passage, it would be premature in the extreme to conclude that temporal passage will not show up in any description of our world in the language of the physical sciences. So the argument from physics is unlikely to dissuade anyone who thinks that because their phenomenology is as of passage, they have good grounds to suppose that there is temporal passage. For given that there are physical models of our world that posit temporal passage, the defender of the argument from temporal phenomenology has good enough grounds to reject [P1]. Or so it seems.

Even if one grants [P1], however, one could still resist the argument from physics by rejecting the claim that the only way to make any difference to our phenomenology is to make a physical difference. To claim that [P4] is true is to claim that physicalism is true: that the only properties are physical properties. Dualists reject such a claim. They think that there exist non-physical properties and they think that these properties make a difference to our phenomenology. In particular, dualists think that some mental properties, such as the way it feels for something to look red, or the way it feels to be in pain, are themselves non-physical properties. So one way of changing the way things seem to us is to change these non-physical mental properties. If we can do that without changing any physical properties, then we can make a difference to our phenomenology without making any physical difference. So dualists can resist this argument by holding that our temporal phenomenology is non-physical, and so it can vary without anything physical in the world varying.

Is this a plausible response on the part of the dualist? Dualists can (and perhaps do) think that the phenomenology of temporal passage is a non-physical mental property. *Our* dualist, however, not only has to say that the feeling of temporal passage is non-physical, but also that *temporal passage itself* is non-physical (since it makes no physical difference). But nothing about dualism about the *mental* makes it plausible that temporal passage itself is non-physical, or that if it is non-physical, it can make a difference to our phenomenology. After all, although dualists think that to have a phenomenology of pain is to be in a non-physical mental state, they think that being, say, stabbed by a dagger, causes that state. Daggers are physical. The stabbing causes (physical) body injury, and causes (physical) brain states, and there being these brain states is part of why we have the non-physical feeling of pain. Our dualist has to say that temporal passage (unlike daggers) is non-physical and makes *no* physical changes to our brains (unlike being stabbed by a dagger) and that *without* any changes to our brains, the presence of this non-physical passage makes a difference to our non-physical mental states. To think this goes well beyond anything to which the dualist is usually committed.

Let us move on now to consider the second argument from physics. That argument begins by noting that it is nomologically possible – this means something like consistent with the actual laws of nature – that there is a world in which one temporal end of the universe is a mirror image of the opposite end. Let's call that world W. W has two boundaries: call them B and B*. Viewed from B, the universe begins, at B, with a big bang, then continues to expand until time t. After t, W begins to contract towards a big crunch at B*. Viewed from B*, the universe begins with a big bang, then continues to expand until time t, then after t W begins to contract towards a big crunch at B (an explosion like the big bang is just a big crunch when viewed from the opposite orientation). Moreover, the portion of W that runs from B until t, is a qualitative duplicate of W as it runs from B* till t. So the moment just after the 'big bang' at B is a qualitative duplicate of the moment just before the 'big crunch' at B* and so on. See Figure 7 for a diagram of a simple world like this.

Suppose that there are persons in W. For any individual, such as Fred, that individual has a Doppelganger at the other end of W whose life looks just the same except in reverse. We could call such a pair of Doppelgangers Fred and Derf. Consider some particular moment of Fred's life. Now consider the duplicate moment of Derf's life. One is a physical duplicate of the other. If physicalism about the mental is true, then the way things seem to Fred at that time is the very same way they seem to Derf at that time. But then, since Derf's life is a physical duplicate of Fred's life, Derf will have the very same experiences as Fred does. Yet if time actually passes, then it must also pass in W. After all, W shares the same laws as the actual world. But if time passes in W, then it makes no difference to the phenomenology of those in W. For Fred and Derf share the same temporal phenomenology, since things seem the same to them. But only one of them is oriented in the direction that time is passing. Time either passes from B to B*, or from B* to B. Whichever way around time passes, its passing makes no difference to temporal phenomenology in W. So we should conclude that it makes no difference to temporal phenomenology in our world either, since our world shares the very same laws as W.

How might one resist the Doppelganger argument? One option is to resist the claim that Fred and Derf have the same experiences. This can

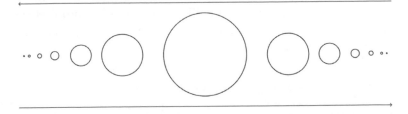

Figure 7 A Mirror World: World W, from big bang to crunch. A mirror world in which the two boundary conditions of the universe feature a singularity.

be achieved by denying physicalism. For then one can say that temporal passage plays some role in determining the phenomenal character of Fred and Derf's experiences, so that even though one is a physical duplicate of the other, they are not phenomenal duplicates. Things seem differently to Derf than they do to Fred.

A second option is to deny that W is nomologically possible. Why think W is nomologically possible? Presumably, because, on the one hand, our best current physics is consistent with W, and, on the other hand, completed physics will be similar enough to current physics that the physical laws will not rule out W. The problem, however, is that if one thinks that actually there is temporal passage, then one probably already thinks that current physics leaves something out: after all, temporal passage does not show up in current physics. So one will think that there is some feature of the physical situation that we haven't captured yet, which makes it the case that the experiences of Fred and Derf do, in fact, differ. After all, only one of these individuals has experiences in the same direction as the passage of time. Indeed, if Fred's experiences 'line-up' with the direction of passage, then one might deny that Derf has any experiences at all, or one might accept that Derf has experiences, but reject the claim that they are the same as Fred's, since Fred and Derf's situations are physically different.

That being said, in so far as one thinks that current best physics is a good guide to the way our world is, one should find both arguments from physics compelling. One should take the fact that temporal passage does not show up in these theories as good evidence that if there were temporal passage, it would make no difference to how things seem to us; better yet, one should probably take this as good evidence that there is no temporal passage. On the other hand, in so far as one is sceptical that current best physics is a good guide to at least some ways the world is, one might not find either argument from physics compelling.

In what follows we consider some other ways one might argue for the falsity of the Difference Making Thesis, by arguing that even if there were temporal passage it would make no difference to our temporal phenomenology. These arguments, however, appeal not to empirical considerations but to metaphysical considerations.

3.7.2. *Arguments from Metaphysics*

If the argument from temporal phenomenology is sound, then it must be that the presence of temporal passage *makes a difference* to our temporal phenomenology. But what does making a difference consist in? One obvious suggestion is that what it is for temporal passage to make a difference to our temporal phenomenology is for it to be the case that had there been no temporal passage, our temporal phenomenology would have been different: in particular, it would not have had the phenomenal content it in fact has. This is a natural thought, since the supposition is that it seems to us that time passes because time passes. So if time did not pass, then it would not seem to us as though time passes. To understand difference making in this way is to understand it in terms of a particular *counterfactual*. A counterfactual is a conditional about what would have happened, had things gone differently from how they in fact went. So if actually Sara had tofu for dinner, it might still be true that had she instead had beef, she would have had food poisoning (if the beef was off). We evaluate counterfactuals by imagining what happens at close possible worlds: worlds that are a lot like our world, except that they differ in small respects. In this particular case we look to a world that is very much like ours except that instead of eating tofu for dinner Sara eats beef, and we look to see whether in that world she contracts food poisoning. If she does, then the counterfactual is true.

If the right way to understand difference making in this context is in terms of counterfactuals, then we will say that there actually being temporal passage makes a difference to our temporal phenomenology just in case in the closest world in which there is no temporal passage, our temporal phenomenology is different (i.e. it does not have a phenomenal content as of time passing). To do so is to understand difference making modally. There is clearly something right about this idea. The problem lies in the fact that many philosophers think that whichever theory of temporal ontology is true, that theory is true in every world. So if our world contains temporal passage, then according to these philosophers there are no worlds with time in which time does not pass. But then any world that lacks temporal passage will be a world that lacks time. Yet it is not clear that we want to draw conclusions about the difference making properties of temporal passage by considering what things seem like in a world *with no time at all*.

Another option for making sense of the difference making in question is to look at what things are like at actual times that are not present. After all, many models according to which time passes are ones in which only some events are present, and which events those are, changes. Since it is the movement of presentness that constitutes temporal passage, we should expect the instantiation of presentness, by some time, to make a difference to how things seem at that time. In that case we might understand difference making not as modal difference making, but instead as temporal difference making. In which case we will say that temporal passage makes a difference to our temporal phenomenology just in case the phenomenal character of our temporal phenomenology at a time, when that time is present, is different from the phenomenal character of our temporal phenomenology when that time is not present.

Which way of understanding difference making you think is the right one – modal or temporal – will depend on whether you think there are possible worlds with time that lack temporal passage, and whether you think there are, actually, times that lack presentness. Presentists will deny the second claim, though growing block theorists and moving spotlight theorists will not; many temporal dynamists will deny the first claim, namely those who think that temporal dynamism is true of necessity.

Given all this, it is not straightforward to answer the question of whether we should think that temporal passage makes a difference to our temporal phenomenology. In what follows we focus on temporal difference making since it is, at least, clear that some temporal dynamists can understand difference making in this manner. Since temporal difference making is only appropriate if one is not a presentist – for it requires that we can ask what things are like at non-present times, and presentists deny that there are any such times – in what follows we will use the moving spotlight theory as our stalking horse for highlighting some reasons why one might be sceptical that temporal passage makes a difference to our temporal phenomenology. Much of what we say will be true if instead one considers a growing block version of temporal dynamism (or some other dynamic theory of time).

First, we need to distinguish *qualitative* from *non-qualitative* properties. Qualitative properties are properties like having a certain mass, shape, colour or texture. Their instantiation makes some in principle observable, measurable or experiential difference to how things are. By contrast, non-qualitative properties are properties like *being Fred*. So, for instance, imagine a possible world containing two individuals who are qualitatively identical. One individual, however, has the non-qualitative property of being Fred, and the other does not have this property. The instantiation of that property makes a difference to how things are in that world: it makes it the case that one of those individuals is Fred, and the other is not. But that's not an observable or experiential difference. No one looking at or poking or prodding the two individuals could determine which of them is Fred, including Fred himself.

So suppose that presentness is a non-qualitative property. Its instantiation, by a time, makes it the case that that time is present, just like the instantiation of being Fred by some individual makes that individual Fred. Then that is the *only* difference instantiating presentness makes. If that is what presentness is like then its arrival makes no qualitative difference to the time at which it arrives, and its departure makes no qualitative difference to the time from which it departs. If that is what presentness is like then it makes no difference to our temporal phenomenology, because it makes no qualitative difference whatsoever, and in order to make a difference to our phenomenology presentness would need to make a qualitative difference: a difference in how things *seem*. So if presentness is a non-qualitative property then its instantiation and movement makes no difference to our temporal phenomenology. If so, the Difference Making Thesis is false.

Suppose, instead, that presentness is a qualitative property. Then it might be that the presence of presentness makes a difference to our temporal phenomenology. Perhaps when our experiences have the light of presentness shining on them they have a different phenomenal character than when they do not have presentness shining on them. The problem, however, is that on at least some dynamic views, such as the moving spotlight view, presentness is not causally efficacious. Let's briefly recap the moving spotlight view for a moment. According to that view, past, present and future objects, properties and events exist. There exist past, present and future times, with presentness 'moving across' the entire temporal span of the universe. Facts about which events cause which other events are already 'fixed' in virtue of past, present and future objects, properties and relations existing. When the very first moment of time is objectively present, it is true that Fred's eating salt some 200 years later causes him to be thirsty. Similarly, when the light of presentness shines on Cleopatra, it is still true that in 5,000 years a colony will be built on the moon and so on. Now, presentness arrives at a time and lights it up, but after presentness has left that time, the time goes back to being just as it was before presentness arrived. Presentness doesn't leave any record of itself. Presentness *itself* does not cause anything.

Think about all of the things in the world with which we are familiar. They are all causally efficacious. We see them: they cause light rays to bounce off them in certain ways. We run into them (or walk around them) because they are impenetrable to matter such as ours; they leave muddy paw prints on our clothes, and hairs on our sofas; they leave traces in our memories, and so on and so forth. Presentness is not like this. It cannot be like this because all of the causal relations are already fixed in virtue of past, present and future moments already being set before presentness ever arrives. So there is no scope for presentness to leave a record of itself at any time. Presentness lights up the time that instantiates it. But that time is, qualitatively, the very same way after presentness leaves it as it was before presentness arrived at it.

Perhaps not all dynamic theories need suppose presentness to be causally inefficacious (we leave it to the reader to think about whether they do, and if not, whether some similar sorts of worries might nevertheless arise for such views). But if presentness is causally inefficacious is its instantiation a good explanation for our having the temporal phenomenology we do? It is hard to see how. After all, we want it to be the case that our temporal phenomenology has the phenomenal content it does *because* of the instantiation and movement of presentness. But we have already seen that according to most philosophical theories of content, mental states get to represent P by bearing some causal connection to P. So if presentness is causally inefficacious then its instantiation does not causally interact with any of our mental states. Thus, on the assumption that our mental states can only represent something if they are (sometimes) causally in contact with that thing, it cannot be that the instantiation of presentness explains our having the temporal phenomenology we do. So we have reasons from metaphysics to think that even if there is temporal passage, it does not make a difference to our temporal phenomenology. If that is right, then the Difference Making Thesis is false, as is [P2] of the argument from temporal phenomenology. The presence of temporal passage does not provide the only reasonable explanation for our having the phenomenology we do, because it fails to provide even a good explanation.

3.8. Summary

In this chapter we have looked in some detail at a particular argument in favour of the dynamic theory of time. This argument focuses on experience. The basic idea is this: it seems to us as though time passes and so we should believe that time really does pass. We then looked at some of the available responses to this kind of argument. The broad debate can be summarised as follows:

(1) The argument from temporal phenomenology seeks to establish the existence of temporal passage based on our having experiences that seem to us to have a certain content.

(2) The cognitive error theorist maintains that our belief that it seems to us as though time passes – our belief that our phenomenology has a certain phenomenal content – is a false belief that arises either because we misdescribe our phenomenology or because we mistakenly infer that it has certain content when it does not.

(3) No content theorists maintain that our temporal phenomenology lacks content altogether, and so it is just not true that it seems to us as though time is passing.

(4) Phenomenal illusion theorists maintain that we do have experiences whose content is as of there being temporal passage, but they hold that these experiences are illusory, since there is in fact no temporal passage.

(5) The passage theorist maintains that we experience the passage of time because time does, in fact, pass.

(6) The makes no difference argument against the passage theorist aims to demonstrate that the passage of time makes no physical difference to the world and thus it cannot explain why we have the experiences that we do (whatever those experiences might be).

(7) The makes no difference argument relies on the controversial assumption that the passage of time makes no physical difference to the universe. One can defend this assumption by appealing either to facts about physics or to facts about metaphysics.

3.9. Exercises

i. Try to work out how a presentist or a growing block theorist might explain our temporal phenomenology. Are there any difficulties that you can see with these putative explanations?

ii. Write down all of the features of your present experience of the world that have something to do with time. Do any of these involve temporal passage?

iii. Think about the experience of change. Is there any way that you can see to argue from the experience of change on its own to the existence of temporal passage?

iv. Divide into three groups: the defence, the prosecution and the evaluation. Put your experience of time on trial. Let the defence defend the view that the experience of time provides evidence for the existence of temporal passage. Let the prosecution argue against this view. Let the evaluation determine a winner during your debate by assigning points for good arguments.

v. Consider your experience of space. Are there any differences that you can identify between the experience of space and the experience of time?
vi. Imagine you are trying to develop a scientific experiment to try to determine what constitutes the human experience of time. What might your hypothesis be? How would you go about testing that hypothesis? Design the experiment.

3.10. Glossary of Terms

Dualism
The view according to which the mind is non-physical.

Non-Qualitative Properties
Properties that do not make some in principle observable, measurable or experiential difference to how things are.

Phenomenal Character
The phenomenal character of a phenomenal state is *what* it's like to be in that mental state.

Phenomenal Content
The content of a phenomenal state is what that state represents about the world, i.e. what it says the world is like.

Phenomenal States
A phenomenal state is a mental state, such that there is something that it is like to be in that mental state.

Physicalism
The view that everything that exists is physical, and hence the mind is physical.

Qualitative Properties
Properties that make some in principle observable, measurable or experiential difference to how things are.

3.11. Further Readings

S. Baron, J. Cusbert, M. Farr, M. Kon and K. Miller (2015) 'Temporal Experience, Temporal Passage and the Cognitive Sciences', *Philosophy*

Compass 10 (8): 560–71. An accessible introduction to the argument from temporal phenomenology and the various ways one might respond to the argument.

B. Dainton (2008) 'The Experience of Time and Change', *Philosophy Compass* 3 (4): 619–38. A fairly accessible, but not introductory, overview of the literature on the experience of time (and change).

I. Phillips (2014) 'Experience of and in Time', *Philosophy Compass* 9 (2): 131–44. A fairly accessible, but not introductory, overview of different theories about our experiences in time (rather than of time), and includes discussion of views about the way our experiences are structured by time which are not outlined in this book.

L. A. Paul (2010) 'Temporal Experience', *Journal of Philosophy* 107 (7): 333–59. This is not an introductory paper, but it is one of the earliest papers outlining phenomenal illusionism. It includes discussion of a range of psychological literature, and is quite accessible.

J. Ismael (2011) 'Temporal Experience' in Craig Callender, ed., *The Oxford Handbook of Philosophy of Time* (Oxford University Press). Also not introductory, this chapter presents an accessible acount of why we have the temporal experiences we do, by appealing to a philosophical theory about how we represent the world at different times.

H. Price (1996) *Time's Arrow & Archimedes' Point: New Directions for the Physics of Time* (Oxford University Press). This book-length treatise on time is quite accessible. It includes an overview of the physics of time, alongside various arguments for why temporal passage would make no physical difference.

4

Time and Physics

In the previous chapter we considered one of the central arguments in favour of the dynamic theory of time – the argument from experience. In this chapter, we will consider the central argument against the dynamic theory of time – the argument from the special and general theories of relativity. This is difficult terrain, since some knowledge of our current best physical theories is needed to fully grasp the problem. We won't attempt to outline these theories here. Instead, we will focus on one of the core implications of relativistic mechanics – the relativity of simultaneity – and use that as a touchstone for our discussion.

4.1. Dynamic Time and Relativity

As we have already seen, the precise nature of temporal passage is controversial, but it is common to understand the idea that time passes in terms of a moving now, which presupposes a privileged present. The privileged present is a moment of time that is metaphysically special in some respect. Exactly which moment is special, changes, and its changing constitutes the flow of time. So, for instance, suppose that 2pm is the metaphysically special moment. Soon, 3pm will become metaphysically special and 2pm will lose its special status. To 'become present' just is to become metaphysically special in this way. Thus, as 2pm loses its special status, it becomes past. Before 3pm gained its special status, it was future. We can thus understand 'past' and 'future' in terms of the privileged present: X is past when it is earlier than the present and X is future when it is later than the present. What it means to be 'metaphysically special' depends on the particular dynamic theory one is focusing on. For instance, according to the moving spotlight theory of time, being 'metaphysically special' means possessing a monadic property of presentness. According to presentism, by contrast, being 'metaphysically special' means being the only moment that exists, and so on.

All dynamic theories of time face empirical refutation (i.e. they risk being shown to be false in the actual world); for there is a tension between the special theory of relativity and the existence of a privileged present. The problem concerns simultaneity. In particular, distant simultaneity: simultaneity between spatially separated events. The special theory of relativity implies that simultaneity is *relative*. As one alters one's inertial frame of reference (a way of assigning spatial and temporal coordinates to events based on a state of constant motion) by speeding up to some constant velocity or slowing down to some constant velocity, what is simultaneous with what, changes. So while X and Y are observed to be simultaneous for one observer, O_1, travelling at half the speed of light, X and Y are observed not to be simultaneous for another observer, O_2, travelling at three-quarters of the speed of light.

The relativity of simultaneity is one of the strangest features of our universe. To gain a feel for this implication of relativity it is useful to consider a thought experiment. Suppose that a train rushes by a platform at half the speed of light. There are two observers: Sara, who is standing on the platform as the train rushes by, and Suzy, who is on the train. Suppose that the two ends of the train are struck by lightning. Suppose that Sara is equidistant between the two lightning strikes. In her frame of reference, the light from the two lightning strikes reaches her together. 'Aha!' She says 'The two lightning strikes are simultaneous!' What does Suzy see? Well, suppose that Suzy is standing in the middle of the train. Because Suzy is on the train, and the train is travelling towards the lightning strike that hit the front of the train and away from the lightning strike that hit the back of the train, Suzy is able to chase down the light that comes from the forward lightning strike. She also runs away from the light from the rear lightning strike. Accordingly, she will see the lightning strike that hit the front of the train before the strike that hit the back of the train. 'Aha!' she will say, 'the forward lightning strike occurred before the backward lightning strike!'

Now here's the rub. There is no physical basis upon which we can say that Sara is correct and that Suzy is wrong or vice versa. Which is to say, there is nothing in our best physics, or in the laws of nature that we have discovered, that would suggest that there is anything special about Sara's perspective, or about Suzy's perspective. As far as we know, both perspectives on the universe – both frames of reference – are equally good. But if that's right, then it seems we are forced to say that whether or not two events are simultaneous depends on the relative state of motion of an observer. In one frame of reference, two events may be simultaneous, but in another frame of reference they may not.

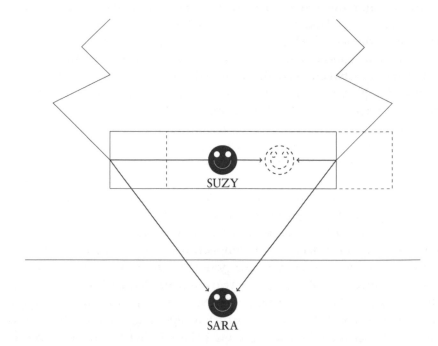

Figure 8 The Train Experiment: Sara and Suzy disagree about the simultaneity of lightning strikes that strike the two ends of the train.

The relativity of simultaneity has radical implications. For what it really shows is that the temporal ordering of events in B-series terms – which events are earlier than, later than or simultaneous with which other events – depends on the perspective of an observer in the universe. There is not just *one* B-series ordering of events. There are many, where each such ordering is indexed to an inertial frame of reference. So even B-theorists have to somewhat amend their theory to accommodate the relativity of simultaneity. But they can do so relatively straightforwardly, by simply saying that there are multiple B-series orderings of events rather than a single, preferred ordering. Since nothing in the B-theory requires such a preferred ordering, the theory is left intact. That, of course, is not so for the A-theorist, who requires that there is such a preferred ordering, as we will see shortly.

Before we turn to consider why the relativity of simultaneity is problematic for dynamic theories of time, it is worth clarifying the extent to which the relativity of simultaneity is, and is not, revisionary of what we might think of as our pre-theoretic view of time. According to the theory that simultaneity is relative, there are events (let's call them E1 and E2) such that observers in one inertial frame will observe E1 and E2 as simultaneous, observers in another inertial frame will observe E1 as occurring before E2 (will observe E1 as *earlier* than E2), and observers in another inertial frame will observe E2 as occurring before E1 (will observe E1 as *later* than E2).

Given this, one might be tempted to think that if simultaneity is relative, then for any pair of events (let's call them E and E*) there is no fact of the matter regarding whether E and E* are simultaneous, or whether E is earlier than E*, or whether E* is earlier than E. That would indeed be a very surprising outcome. It would mean that there will be observers who observe the following sequence of events: Dino (the dinosaur) eats some food, then Dino walks through the forest, then Dino is eaten by a larger dinosaur; and there will be other observers who observe the following sequence of events: Dino walks through the forest, then Dino is eaten by a larger dinosaur, then Dino eats some food.

It is important to see that this is not the case, and to see why. Observers in different inertial frames will disagree about the temporal order of *certain* events: namely, all those events that are space-like separated from one another. These are events which light cannot travel between. Or, to put the point another way, if one wanted to travel between these events, one would need to travel faster than the speed of light to do so. For these events, there is simply no fact of the matter regarding their temporal order: relative to some inertial frames these events will be observed to be simultaneous, and relative to others, will be observed to be in different temporal orders (one earlier than the other, or one later than the other).

Not all events, however, are like this. Consider Dino who walks through the forest, runs into a large dinosaur, and is then eaten. These events are causally connected (or at least, they might be). Causally connected events, or events that *can* be causally connected, are such that a sub-luminal signal can travel between these events: that is, a signal travelling at less than the speed of light can travel between these events. These are called time-like separated events. Such events are always observed, regardless of the observer's inertial frame, to be in the same temporal order. So, returning to Dino, all observers, in every frame of reference, will observe Dino to wander in the forest, to eat, and then to be eaten by the large dinosaur. (More on this in section 4.2.)

What is important to see is that *simultaneity* can be relative without it being the case that *all* temporal order is relative. Events that are time-like separated have an invariant temporal ordering: they will be observed to have the very same temporal ordering in every frame of reference. These events will *never* be observed to be simultaneous with one another. It is only events that are *not* time-like separated, but instead are space-like separated, which will be observed to be simultaneous by observers in some inertial frames, and observed to be non-simultaneous by observers in other inertial frames.

Having clarified these issues, we can now return to consider the connection between the relativity of simultaneity and dynamic theories of time. It is relatively easy to see why the relativity of simultaneity will create problems for any dynamic theory of time. For it is natural to think that *all* of the events that are metaphysically special, by being present, are simultaneous with one another, and that none of those events is simultaneous with any event that is *not* present. That's because it is, in turn, natural to suppose that if there is a metaphysically special present, then all the events that are not present (if there are any) are either objectively future, or objectively past.

We can capture this idea via the following principle:

[P2S] For any X and Y, if X is present then Y is present if and only if X is simultaneous with Y.

[P2S] forges a link between the dynamic theory's concept of presentness and the concept of simultaneity, providing a definition of what it is for any X and Y to occupy the same privileged present. [P2S] tells us that if X is present, then Y is also present only if X and Y are simultaneous; it also tells us that if X is present, then Y is present if X is simultaneous with Y.

In order to state the conflict between the dynamic theory of time and the relativity of simultaneity a bit more set-up is required. First, it is useful to define the notion of a simultaneity class as follows:

[Simultaneity Class] Set C is a simultaneity class iff all members of C are simultaneous with one another.

So, in a simple world where simultaneity is not relative, and which contains only three events, E1, E2 and E3, and E2 and E3 are simultaneous with one another (and not with E1) then one simultaneity class in that world contains just E2 and E3, and the other contains just E1. A simultaneity class can be (very roughly) thought of as an instant of time. This is not quite right in a relativistic setting, but the simplification won't cause problems in what follows. Thus, if E2 and E3 are simultaneous with one another and E1 is not simultaneous with anything in the imagined scenario, then E2 and E3 occur at a different moment of time to E1 (see Figure 9).

Next, we need a particular case to focus on. Consider an observer O_1 sitting in their office at 2pm on Thursday 25th May 2017. At 2pm, O_1 looks up and sees a second observer, O_2, pass their window at half the speed of light (see Figure 10).

Since simultaneity is relative, in the example just outlined there are two simultaneity classes: the class defined relative to O_1's frame of reference, and the class defined relative to O_2's frame of reference. Now, reason as follows. Suppose that O_1 is in the privileged present. At 2pm, O_2 is inside O_1's simultaneity class, C_1. By the relativity of simultaneity, what is simultaneous for O_2 is distinct from what is simultaneous for O_1. Thus, O_2 is also

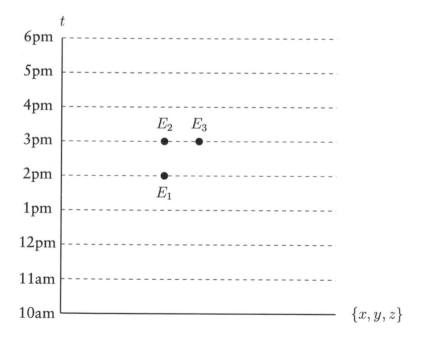

Figure 9 Simultaneity: Three events occurring at different times. E2 and E3 are simultaneous with one another. E1 is not simultaneous with either event. E2 and E3 are part of one simultaneity class, E1 is part of another.

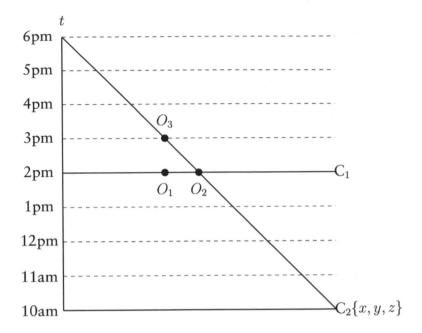

Figure 10 Relative Simultaneity: Two observers in relative motion will disagree about what is simultaneous with what. C_1 corresponds to O_1's simultaneity class; C_2 corresponds to O_2's simultaneity class. The vertical axis is time, the horizontal axis is space.

a member of a distinct simultaneity class, C_2. It follows by the distinctness of C_1 and C_2 that there is something, O_3, that is simultaneous with O_2 but not O_1. But now turn back to our principle, [P2S]. Applying that principle to this case, everything inside C_1 is present (since O_1 is in the present). So O_2 is present since O_2 is in C_1. Applying [P2S] again, though, everything in C_2 is present, since everything in C_1 is simultaneous with O_2. So O_3 is present. But now consider O_1 and O_3: according to our reasoning so far they are *both* present, and yet O_1 is not simultaneous with O_3, which contradicts [P2S].

An analogous argument reveals that there are things for O_2 that are present but that are not simultaneous with that observer. Indeed, for any observer the same argument can be generated. Moreover, there is nothing special about observers. We could easily formulate the argument in terms of particles moving relative to one another. All that we really need to get the problem going is the idea that there are different inertial frames of reference and that, across those frames of reference, the distant simultaneity between events changes.

Because a contradiction is derived, one of our assumptions has to go: either we reject the relativity of simultaneity, or we reject [P2S], or we reject the claim that there is a privileged present. If we *assume* [P2S] and the relativity of simultaneity, then an argument against the dynamic theory of time may be set out as follows:

The Argument from Simultaneity

[A1] The dynamic theory of time is true.

[A2] [P2S] is true.

[A3] Simultaneity is relative to an inertial frame of reference.

[A4] If the dynamic theory of time is true then there is a privileged present.

[A5] If there is a privileged present and simultaneity is relative to an inertial frame of reference, then [P2S] is false.

Therefore,

[C] The dynamic theory of time is false.

Note that [A3] is underpinned by a second, sub-argument: [B]. This argument aims to establish that the relativity of simultaneity is a real phenomenon. It will be useful in what follows to have this sub-argument to refer back to, and so we set it out as follows:

Sub-Argument B

[B1] If the special theory of relativity is true, then simultaneity is relative to an inertial frame of reference.

[B2] The special theory of relativity is true.

Therefore,

[C] Simultaneity is relative to an inertial frame of reference.

A dynamic theorist has three responses available to the argument from simultaneity. First, she can deny [A2] by giving up or weakening [P2S]. Responses of this kind will be considered in section 4.2. Second, she can deny [A4] and thus deny the claim that the dynamic theory of time is committed to a privileged present. We will turn to this class of responses in section 4.3. Finally, she can reject [A3] by taking issue with sub-argument [B]. This line of response is taken up in section 4.4.

4.2. Dynamic Time and Simultaneity

As noted, the first response available to the dynamic theorist is to deny [P2S]. [P2S], as already discussed, forges a link between the notion of a privileged present and the concept of simultaneity. This link is useful because it tells us what it means for two events to be present with one another or 'co-present'. Without [P2S] it is unclear what it means to say that two events are co-present. Accordingly, if the dynamic theorist rejects [P2S] something must be put in its place. A new definition of what it is for any X and Y to be present together must be offered, one that does not yield a contradiction when combined with the relativity of simultaneity. But what would such a definition look like? To answer this question, dynamic theorists look to the geometrical structure of spacetime.

We ordinarily think of space and time in a certain way: each time corresponds to a complete specification of the spatial location of every object in the universe at that time. Moreover, there is a single, total ordering of times so construed. In a relativistic setting, however, this way of thinking about space and time can no longer be sustained. There is no single, total ordering of times (at least for spatially separated events, as already noted). Rather, there are many such orderings, each of which is indexed to a particular inertial frame of reference.

We can imagine the difference, roughly, by thinking about stacks of shapes. Suppose we represent a time as a piece of paper. The area of the sheet represents the entirety of space at that moment in time. In a pre-relativistic setting, the universe is a single stack of sheets each of which is the same shape (see Figure 11).

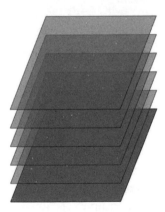

Figure 11 Sheets of Time: The universe as a sequence of times.

In a relativistic setting, however, we have papers of different shapes. If we take two opposite edges of the paper, then as we increase our motion, the angles of those edges change. This change in the angle of the sheet of paper represents the change in temporal distance between events. The more that the angle of the sheet is skewed, the greater the change to the temporal distance between spatially separated events. Moreover, in a relativistic setting, the universe is not a *single* stack of sheets of the same shape. Rather, while we can represent the universe as a stack of sheets of the same shape within a frame of reference, there is no one stack that we can use to represent the entire universe regardless of the frame of reference. Each frame corresponds to a different stack (see Figure 12).

Minkowski spacetime is a way of 'weaving' all of the stacks together to produce a super-stack that *does* represent the entire universe. Doing this requires an entirely new geometry: a four-dimensional geometry that is capable of combining the various stacks of shapes into a single stack which has a single shape; a shape that is utterly unlike any of the shapes of the individual stacks. In this geometry space and time are woven together into a single metric, where a 'metric' is (very roughly) just a way of determining the distances between things. Locations in the universe are spacetime points (as opposed to spatial locations or times). A spacetime point can be dismantled into a spatial and a temporal location, in a frame of reference. Which is to say that within a frame of reference we can unpick a spacetime point to get us back to a sheet of paper of a particular shape, though which shape that is will depend on how fast one is going at the relevant location.

Minkowski spacetime has a number of invariant features; features that do not depend upon one's inertial frame of reference. In particular, while spatial and temporal distances in relativity depend upon one's relative state of motion, spatiotemporal distances within Minkowski spacetime do not. Minkowski spacetime therefore provides an invariant metric for describing the distances between things.

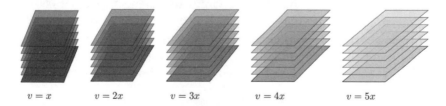

$v = x$ $v = 2x$ $v = 3x$ $v = 4x$ $v = 5x$

Figure 12 Velocity Transformations: The universe as a sequence of times by reference frame. V is velocity. Different velocities correspond to different inertial frames. Let x = 50,000 kph.

To understand the basic idea behind Minkowski spacetime as well as the reason why a new geometry is required at all, it is useful to consider an ordinary Euclidean geometry for space. Euclidean geometry is the geometry that most of us learn at high-school. A Euclidean geometry for space uses a simple metric. In Euclidean geometry, the distance between any two points A and B in a three-dimensional space is given as follows: $\Delta d^2 = \Delta x^2 + \Delta y^2 + \Delta z^2$. In other words, the distance between A and B is determined by combining the distances between A and B along the x, y and z axes to come up with a total distance. For instance, suppose that one sets up a coordinate system and, within that coordinate system, A's (x, y, z) coordinates are: (2, 3, 4) whereas B's (x, y, z) coordinates are: (–2, –3, –4). Then the distance between these two points is:

$$\Delta d^2 = (2+2)^2 + (3+3)^2 + (4+4)^2$$
$$\Delta d^2 = 4^2 + 6^2 + 8^2$$
$$\Delta d^2 = 16 + 36 + 64$$
$$\Delta d^2 = 16 + 36 + 64$$
$$\Delta d^2 = 116$$
$$\Delta d = \sqrt{116}$$
$$\Delta d = 10.77$$

A Euclidean geometry is *invariant* under transformation of the base coordinate system. First, what is a transformation of coordinate systems? Well, here's an example. Consider Sara. Sara, an Earthling, wants to know the distance between two antipodal points on the Earth's surface: New York, USA, and a location just south-west of Perth, Western Australia (let's just say Perth for the sake of argument). She determines the distance as follows. First, she situates both Perth and New York within a coordinate system. To do this, she sets some point in space to be the origin of the x, y and z axes. All this means is that the origin of the Earth has the coordinates (0, 0, 0). She then assigns x, y and z coordinates to everything in space based around that origin. So, suppose that in her coordinate system, 1 unit = 1,000 km. She determines that Perth's location is (0,7,0) and New York is located at (0,–7,0). In short, being antipodal points, there is only *one* axis along which the two points differ from the origin at the centre of the Earth. She then calculates the distance between the two using the above equation as follows:

$$\Delta d^2 = 0^2 + 14^2 + 0^2$$
$$\Delta d^2 = \sqrt{196}$$
$$\Delta d = 14$$

Sara thus concludes that Perth and New York are 14,000 km apart if one were to tunnel straight through the Earth from one side to the other.

Now, we can transform this coordinate system simply by moving the origin, whilst keeping the distance between each point in the x, y and z axes fixed. Thus, consider Zara, who lives on the moon. She wants to work out the distance between New York and Perth as well. Zara, however, is extremely fond of her home and can't think of doing anything but setting the origin of her coordinate system at the centre of the moon. Even if she moves the point of origin in this way, Perth and New York must still retain their distances from one another along each axis. Thus, let us suppose for simplicity that Perth is located at (384,384,384) for Zara. Then New York will be located at (384,370,384), and so the calculated distance between Perth and New York for Zara will be the same for her as for Sara.

The fact that Zara and Sara can agree on the distance between New York and Perth is extremely useful for both of them. It allows coordination about a number of things, but most importantly it allows both Zara and Sara to produce empirically equivalent accounts of the world *despite* their relative differences in perspectives, and thus despite the way in which they produce a coordinate system for determining the distances between events. Why does that matter? Well, suppose that Sara and Zara want to collaborate to send a probe to the centre of the Earth half way between Perth and New York. To do this, they need to be able to speak a common language about distance, location and so on, otherwise they have no hope of coordinating the project. The fact that one can transform freely between the two geometric descriptions they provide of the world while still agreeing on all of the relevant facts makes this kind of coordination possible.

The Euclidean geometry just described is purely spatial; there is no mention of time. We can add time into the picture by adding a temporal parameter. We do this by giving everyone a single parameter, in which events are assigned a location in time in exactly the same way for everyone. At each time, each object has a location in space. This location can, however, change. So, for Sara, at each moment of time, she can locate each object in her Euclidean geometry that takes the centre of the Earth as its origin. Each time for Sara is thus like a three-dimensional picture of the universe in which every object is located somewhere in her coordinate system. Zara can do exactly the same thing from her coordinate system. Importantly, Sara and Zara will agree on the temporal and spatial distances between all objects, at all times. Thus, they will agree on how fast everything is going, on when events occur and which things are not moving at all.

The picture just described in which we take Euclidean geometry and insert a time parameter corresponds roughly to our picture of the universe *before* relativity. It corresponds to our first picture of stacked sheets used above. After relativity, this picture won't do, as we have already seen. There are many stacks, and they all have different shapes. To say that the stacks are different shapes is to say that the temporal distance between events is not invariant for different observers. Depending on where you are and how fast you are going, the time between events shifts accordingly. But, and although we haven't really discussed this point, it is also the case that the *spatial* distance between objects fails to be invariant. This can be shown using a variant of the train case described above (though we won't go into the details here, just take our word for it).

For these reasons a Euclidean geometry using a single time parameter won't work for relativity. It can't handle the variance in time and space that we know to exist. So if Sara and Zara use a Euclidean geometry with a single time parameter, they have little hope of agreeing on anything: coordination would be hopeless. Still, the motivations for having some kind of invariant geometry remain. We want an invariant geometry for describing the universe so that different observers can produce empirically equivalent descriptions of the world; descriptions of the world that don't depend on who the observer is. Of course, the Sara and Zara case is a bit artificial. So here's an example closer to home. Consider your phone, which is on Earth, and a satellite. These objects occupy different frames of reference and thus different coordinate systems that carve up space and time in different ways. To get your phone and the satellite to talk to each other, however, there needs to be a common geometrical language that the two objects can speak. If there isn't, then the two objects won't be able to agree with one another, and syncing a phone to a satellite simply won't be possible. The need to get the GPS on your phone to work *at all* is what partially motivates the production of an invariant way to describe the locations of things, a way that rises above the fact that spatial and temporal distances between things change depending on a coordinate system.

But while it is not the case that spatial distance between things is invariant between coordinate systems, it is the case that the combined *spatial and temporal* distance between things is invariant. This means that while two observers will disagree on how far apart in space two things are, or how far apart in time two things are, the *combined* distance between two things in space *and* time will engender agreement. Minkowski spacetime is essentially a geometry that captures this fact. Instead of treating space and time as two separate things – one to be captured by Euclidean geometry, and one to be treated as an external parameter in which 'slices' of Euclidean geometry can be situated – we treat space and time together in a single geometry. The metric is thus not a purely spatial metric, as in the Euclidean case. Rather, the metric is a spatiotemporal metric. The metric is invariant between observers and so can be used to get your phone and a satellite to talk to one another (this is exactly what relativity is used for!).

The metric for Minkowski spacetime is quite different to the metric for Euclidean space. Like Euclidean space, however, it is possible to state the metric algebraically as follows. For any two points A and B, the *spatiotemporal* interval between those points is:

$$\Delta s^2 = -(\Delta ct)^2 + \Delta x^2 + \Delta y^2 + \Delta z^2$$

Where 'c' is the speed of light, 't' is time and x, y, and z are spatial coordinates.

The Euclidean and Minkowskian metrics are actually quite similar. Indeed, the Minkowskian metric *collapses* into the Euclidean metric in three spatial dimensions, since in that situation the temporal aspect of the metric plays no role. Despite that, there are some important differences between the Euclidean and Minkowskian metrics. First, the metric itself uses the speed of light to determine the spatiotemporal distance between events. As a result, the metric has a 'luminal structure'. To return to our metaphor of 'weaving' shapes together: light constitutes Minkowski's thread.

Second, the metric explicitly makes use of space and time *together*. The Euclidean metric made no use of time at all: it was purely spatial. Because time appears inside the Minkowskian metric itself, there is already a sense in which space and time are not clearly differentiated. Whereas before, time was treated as an extra parameter within which sheets of Euclidean space could be situated, here it is treated as one of the dimensions, which, along with space, allows us to locate events and objects. Time, on this view, is much more like a dimension of space, at least geometrically. That being said, time is treated quite differently in this metric compared to space. Time has a *negative* value, while the spatial values are all positive.

As a direct result of the negative signature of the time value in the metric, there are three possible values for the distances between any two spacetime points: positive, negative and null. These are known as: space-like separation, time-like separation and light-like separation respectively. Compare this to Euclidean geometry where there are only two types of value that the distance between two things can take, namely: null and positive. When two things are *null* separated in Euclidean geometry they are located in the same place, when they are *positively* separated, they are at some spatial distance from one another.

Things are quite different in the Minkowskian case. When two spacetime points are space-like separated, a signal must travel at a speed greater than the speed of light to get from one to the other. When two points are light-like separated a signal must travel at the speed of light to get from one to the other; and when two points are time-like separated a signal may travel below the speed of light to get from one to the other. For any spacetime point, p, the set of spacetime points s_1 that may be reached from p at the speed of light constitute p's *future light cone*. Similarly, the set of spacetime points s_2 that may reach p at the speed of light constitute p's *past light cone*. Each light cone has a surface and an interior. The members of s_1 and the members of s_2 are all light-like separated from p and all fall across the surface of p's light cone. Points that are on the inside of p's future light cone are all of those points that can be reached from p at a speed that is slower than the speed of light. Points that are on the inside of p's past light cone are all of those points from which one can reach p at a speed that is

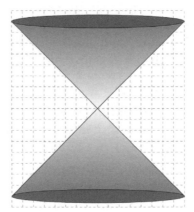

Figure 13 The Luminal Structure of Minkowski Spacetime: The cones correspond to points that are time-like and light-like separated from the central point in the diagram. The top cone corresponds to the future light cone of that central point, the bottom cone corresponds to the past light cone of that central point. The region outside the cones is the absolute elsewhere of the central point. These are space-like separated from the central point. Points that are on the surface of the forward and backward lightcones are light-like separated from the central point. Points that are on the interior of the forward and backward light-cones are time-like separated from the central point.

slower than the speed of light. The set of points s_3 that are all space-like separated from p are called p's *absolute elsewhere* (see Figure 13).

Because the speed of light is a part of the metric of Minkowski spacetime, we can represent the three types of value in Minkowski spacetime using a light cone. The light cone of a spacetime point is a useful way of depicting the three types of distance values that a given spacetime point may bear to others.

We represent *null* values as though we were shining a light onto the points that are at a null spatiotemporal distance from the origin. Imagine turning a torch on: it produces a kind of 'cone' of light that shoots outward, illuminating things as it goes. This is sort of what happens in a light cone diagram. The central point is where we turn the torch on. What gets 'lit up' are all of the points that the light reaches from that central point. The difference, however, is that the torch is not really pointed at anything. Instead, what we are imagining is a globe of light, emanating in space in all directions. We represent the globe as a *double cone* because we are flattening three spatial dimensions into two. Really, we should be drawing a *sphere* of light. But that is a three-dimensional object, and we wouldn't be left with a way to draw time in such a picture!

A double cone, which gives the impression that the light is not *directed* in space, but is emanating in *all* directions, is about the best we can do to represent Minkowski spacetime while still leaving room to depict things that the light cannot reach. These areas – the ones that the light cannot reach – are depicted as lying outside of the cone. Finally, to continue with the metaphor of a torch, the *inside* of the cone represents things that the torch illuminates, but that *could* be illuminated by something going a bit slower than light. So, for instance, suppose we have a sonar as well as a torch. The points on the *inside* of the light cone are ones that a pulse from the sonar can illuminate. Again, this is just a metaphor. We are simply trying to provide a useful way to depict the three kinds of values (+,–,0) in a way that roughly corresponds to the structure of the geometry that the metric expresses, and that can help us to get an intuitive grasp on the picture of reality that the geometry recommends.

With an understanding of Minkowski spacetime in hand, let us now return to the point at hand. Recall that we began looking at Minkowski spacetime because we were looking for a new way to define the present. The reason why Minkowski spacetime might be useful for this purpose is because, as already noted, the spacetime distance between any two points is *invariant*. So if the present is defined in terms of some spatiotemporal distance, then that definition of the present will be immune to any problems that arise due to the *variance* in the temporal distance between things (this, of course, being the problem posed for dynamic theories of time by relativity).

Within the context of Minkowski spacetime several options for defining the present have been identified, each of which provides an account of what it is for any X and Y to be present together. Each definition of the present treats the present as a set of spacetime points. The options in Figure 14 are (arguably) natural candidates that may be used to define the present. But it is open to the proponent of a dynamic theory of time to define the present as *any* set of spacetime points. In effect, what this means is that the dynamic theorist can redefine the present as any spatiotemporal region of the universe that they want.

Assuming that spacetime is continuous (between any two spacetime points there is a third) there is an infinite number of possible ways to define the present, since there are infinitely many constructible sets of spacetime points. So which set of spacetime points provides the best definition of the present? The question is a troubling one, for there doesn't appear to be a non-arbitrary basis for selecting a particular set of spacetime points as the one that underwrites the privileged present.

In Figure 14, everything inside the backwards light cone is in D's absolute past. That means that observers in any inertial frame will observe those events to have a single temporal ordering. Likewise for events in D's future light cone. Only events in the absolute elsewhere of D will be observed, by observers in some inertial frames, as being simultaneous, and by observers in other inertial frames, as being non-simultaneous.

Figure 14 represents the various views one might have about which sets of points count as being present. As we can see, there is some dramatic variation. At one end of the spectrum we have point presentism, according to which only D itself is present. At the other end of the spectrum we have everywhere presentism, according to which the total set of points in the diagram is present. Views part way along the spectrum delineate various different sets of points as being present (see the diagram for the variations here). These various options raise a problem for the response to the argument from simultaneity under consideration; namely, how to decide which set of points to count as present, and, more pressingly, how to defend that choice as the right one. That is, why isn't it simply arbitrary to stipulate that this rather than that set of points is present, given that there are a number of options on the table?

There are two responses to this arbitrariness problem. First, one might argue that no particular set of spacetime points is privileged. Rather *every*

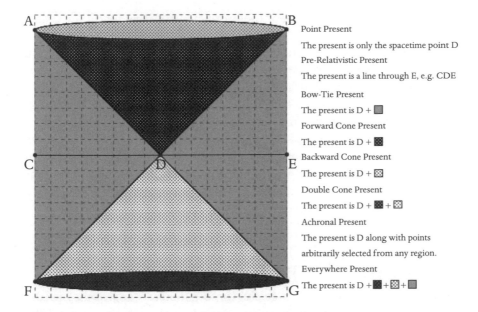

Figure 14 Redefining the Present: Forward, backward and double cone presentism include D plus all of the points that are on the surface of either D's backward or forward light cones (or both). The spacetime points on the interior of D's forward light cone constitute D's absolute future. The points on the interior of D's backward light cone constitute D's absolute past. Any line connecting D with points in D's absolute elsewhere constitutes the simultaneity class of some reference frame.

set of spacetime points defines its own unique, privileged present (e.g. every distinct singleton set). If, however, every spacetime point is present, then it would seem that the entire universe is present. But that seems to undermine the idea that there is anything metaphysically special about the present. If everyone is special, then no one is.

The second solution to the arbitrariness problem takes it as a brute fact that a particular set of spacetime points is present. A brute fact is nothing more than a fact that simply cannot be explained. The thought, then, is to just accept that the present is thus and so, without being willing to answer the question as to *why*. Perhaps the bruteness of this fact is to be expected. After all, consider a dynamic theory of time in a pre-relativistic setting. According to such a theory, exactly one time is present. But which time? And why that time and not some other? Arguably, there is no good answer to this question either. It is a brute fact that *this* time is present.

We shouldn't worry, then, about the brute fact that *these* spacetime points are present.

The analogy between pre-relativistic and relativistic accounts of the present is not perfect, however. Saying that a particular time is present is different to saying that a particular spacetime point is present. As discussed, spacetime points can be dismantled into spatial and temporal locations. Accordingly, when a spacetime point is privileged we are privileging a temporal location *and* a spatial location. We are saying that there is something privileged about the *here* as well as the *now*. The brute fact in the spacetime case, then, is a different type of brute fact. So it is unclear how much we can rely on an analogy to the pre-relativistic case to justify positing the brute fact in question.

4.3. Physical and Metaphysical Privilege

At any rate, the chief challenge for these geometric approaches to the dynamic theory of time is to give some plausible account of why the particular region of the universe that they have carved out as present is, in fact, present. Providing such an account looks difficult. Perhaps, then, it would be better to retain [P2S]. By retaining [P2S] there is no need to redefine the present in a manner that detaches it from simultaneity.

Instead of denying [P2S], the dynamic theorist may seek to weaken it. There are two obvious ways to weaken the principle:

[P2S$_1$] For any X and Y, if X is present then Y is present *if* X is simultaneous with Y.

[P2S$_2$] For any X and Y, if X is present then Y is present *only if* X is simultaneous with Y.

The difference between these principles is just this. [P2S] makes it the case that simultaneity is both necessary and sufficient for two events to be co-present with one another. Thus, when X is present and Y is present, then it follows that X and Y are simultaneous. Moreover, when X is present and Y is simultaneous with X, then Y is also present. The two alternative principles just outlined weaken [P2S] either by denying the necessity of simultaneity for co-presentness or by denying the sufficiency of simultaneity for co-presentness. [P2S$_1$] holds that simultaneity is sufficient but not necessary for co-presentness. [P2S$_2$] holds that simultaneity is necessary but not sufficient for co-presentness.

Only one of these principles allows the dynamic theorist to avoid the argument from simultaneity. [P2S$_1$], in particular, is trouble. For it is natural to suppose that for any X and Y, if X and Y are not simultaneous with one another, then either X is earlier than Y or Y is earlier than X. It is also natural to suppose that if X is earlier than Y and Y is present, then X is past and if X is earlier than Y and X is present, then Y is future. Now, consider again O$_1$ and O$_2$. O$_2$ is in O$_1$'s simultaneity class. So by [P2S$_1$], O$_2$ is present. O$_3$ is in O$_2$'s simultaneity class, so by [P2S$_1$] again, O$_3$ is present. But now consider O$_1$ and O$_3$. O$_1$ and O$_3$ are not simultaneous. So either O$_3$ is earlier than O$_1$ or O$_3$ is later than O$_1$. So either O$_3$ is in O$_1$'s past or O$_1$ is in O$_3$'s past. Either way, O$_3$ is not present (assuming that 'past' and 'present' are mutually exclusive attributions). But both O$_1$ and O$_3$ are present. So we are faced with another contradiction.

[P2S$_2$] is a better bet for the dynamic theorist. That's because this principle cannot be used to generate a contradiction in the same fashion. Given [P2S$_2$], it does not follow from O$_2$'s being in O$_1$'s simultaneity class that O$_2$ is present. All we know is that when O$_2$ is present with O$_1$, then they are in the same simultaneity class. Indeed, even if we can somehow establish that O$_2$ *is* present, it doesn't follow that O$_3$ is present either, even though O$_3$ is simultaneous with O$_2$.

Furthermore, [P2S$_2$] keeps the relationship between presentness and simultaneity intact: because for any X and Y, if X and Y are present together then they are simultaneous, the present corresponds to a particular simultaneity class. Exactly which simultaneity class is present, however, is unclear. But at least there is a conceptual link between presentness and simultaneity. The disadvantage of [P2S$_2$] is that it doesn't tell us what it takes for two entities to be present together. Even if we know that O$_1$ is present, we don't have a way of knowing – based on [P2S$_2$] alone – whether or not O$_2$ is present as well. In order to gain this further piece of information, we'd need to know *which* simultaneity class corresponds to the present.

Now, one might worry that [P2S$_2$] conflicts with the special theory of relativity. As noted above, there is no physical basis for saying that simultaneity for any observer represents the 'true' account of simultaneity for X and Y. To return to the case of Sara and Suzy, there is no sense in which Sara's judgements about what is simultaneous with what are correct, while Suzy's judgements are incorrect, or vice versa. Given [P2S$_2$], however, there is always just one simultaneity class that is privileged in virtue of that class being present. As was just noted, [P2S$_2$] ties the present to a *particular* simultaneity class. Accordingly, one might argue that there is, in fact, a basis for saying that simultaneity for some observer is the correct account of simultaneity. Those events that are in the simultaneity class that corresponds to the present are the only events that are *really* or *truly* simultaneous. Everything else is not *really* simultaneous, because it is not in the present.

This kind of worry rests on a mistake, however. When we say that there is some basis for thinking that Sara's judgements about simultaneity are the correct judgements, or that Suzy's judgements are the correct judgements, the notion of 'correctness' being used is a very specific one. What we mean is that there is something special *from the perspective of physics* with regard either to Sara's perspective or Suzy's perspective; which is to say that there is some empirically discoverable fact about one of these perspectives that would allow us to say that Sara or Suzy is in error.

Consideration of a slightly different case will help to draw the idea out. Suppose that Sara and Suzy are both looking at the sky. Sara, however, is wearing tinted shades, while Suzy is not. 'The sky is blue!' says Suzy. 'No it's not, it's pink!' says Sara. In this case there is a physical basis upon which we can say that Sara is wrong and Suzy is right: Sara's perspective is distorted by her tinted shades, but Suzy's isn't. So there is a physical difference between the two perspectives.

This is what we are looking for in the case of simultaneity: some physically discoverable difference between Sara and Suzy that would allow us to say that one of these perspectives is distorted, while the other is not. Thus, when we say that a simultaneity class is *physically privileged* we mean that there is some physical basis for treating that class differently from other classes. This is standardly understood via the laws of nature. A simultaneity class is physically privileged when the laws of nature behave differently in the frame of reference used to determine the class in question, as compared to the frames of reference used to determine other classes.

This is not, however, the only kind of privilege that might attach to a simultaneity class. When a simultaneity class is *metaphysically privileged* we might say that the class is treated in a unique respect in the context of our best metaphysical theories: i.e. that the class corresponds to the privileged present. Importantly, a simultaneity class might be metaphysically privileged without being physically privileged, so long as the present is not physically detectable. Moreover, although the special theory of relativity casts doubt on the idea that any simultaneity class is physically privileged, it is silent on other kinds of privilege. So it does not rule it out that some particular simultaneity class corresponds to the metaphysically privileged present.

The difference between a physically privileged perspective and a metaphysically privileged perspective can be difficult to grasp. But here's the basic idea. Consider again Sara and Suzy looking at the blue sky. When Sara is wearing her tinted shades, we might say that her perspective is being distorted, and thus that Suzy's perspective is physically preferred. Key to this assessment is the idea that we can discover the difference. When a perspective is metaphysically preferred, the thought is that we cannot discover the difference in the same way. There is nothing physically special about the perspective. Rather, there is some further feature of the perspective that is special, over and above any of the physical facts.

Put this way, the notion of a metaphysically privileged perspective can sound a bit mysterious. After all, we're being asked to believe in some non-physical difference between two points of view. We have no intention of trying to relieve the mystery here. Our goal is only to give the reader a flavour for the kind of view being proposed. At any rate, let us suppose that this notion of metaphysical privilege is sensible. The thought is that by adopting [P2S$_2$] plus the notion of a metaphysically privileged simultaneity class, the dynamic theorist can avoid the argument from simultaneity whilst holding on to the idea that there is a privileged present.

The idea that some simultaneity class is metaphysically, and not physically, privileged gives rise to two difficulties. First, it may not be in the best interests of the dynamic theory of time to cleave metaphysical from physical privilege. As we saw in Chapter 3, passage theorists hold that we have experiences as of time passing, and that we have those experiences because time does in fact pass. Temporal passage just is the movement of a privileged present. So it would stand to reason that if we experience time's passing, it must be because we have some experiential connection to the privileged present.

But if the very thing that makes the present privileged is not physical, then it is quite difficult to see how the passage of time could be being experienced. In short, the idea that the present is metaphysically privileged would render the dynamic theory of time susceptible to the makes no difference argument discussed in section 3.7.

Second, it is troubling that the metaphysically privileged simultaneity class is not physically discoverable, at least when that simultaneity class is being used to delimit the present. On such a view it would seem that nature is 'conspiring' to keep the privileged present hidden from us. This leads to a sceptical conclusion about the present: for all we know, nothing that is simultaneous with us is present. We may not be in the privileged simultaneity class and thus may not have the privileged perspective on the universe needed to be present. On some dynamic theories of time, this has alarming implications. Presentists maintain that only present entities exist. For the presentist, then, we have no way of knowing whether the things we take to be simultaneous with us even exist!

4.4. Dynamic Time and the Privileged Present

So far we have considered responses to the argument from the relativity of simultaneity that involve denying or weakening [P2S]. Next we will look at solutions that involve denying [A4]: the claim that if the dynamic theory of time is true then there is a privileged present. There are two directions we could go: we could multiply the present (there isn't *a* privileged present, there are many) or we could diminish it (there isn't even one privileged present). Let us consider each option in turn.

4.4.1. Multiplying the Present

The most natural way to multiply the present is to index the present to frames of reference. Every frame of reference is used to specify a particular simultaneity class, which is then taken to be the present for that frame. Having indexed the present in this manner, there are then two different ways to develop the dynamic theory of time: fragmentalism and unificationism. The *fragmentalist* maintains that, from the perspective of each reference frame, the privileged present associated with other frames doesn't exist. The *unificationist* disagrees: from the perspective of each reference frame, the privileged present associated with other frames exists.

We can clarify the distinction by thinking of the world from a 'god's eye' perspective. On both views, if one were to take a 'god's eye' perspective on the universe, one would see that there are many privileged presents. The fragmentalist, however, will use this notion of a 'god's eye' perspective as a heuristic device only. For the fragmentalist, there is no such perspective to be had, at least not one with any metaphysical significance. For the unificationist, by contrast, the 'god's eye' perspective is metaphysically apt. Fragmentalism thus involves giving up the idea that there is a globally consistent metaphysical description of the universe. The universe is divided into a number of distinct 'realities', where each such reality displays a distinct privileged present. The extent to which this is plausible really depends on the degree to which one can make sense of a reality divided.

On either view, [P2S] must be sharpened. [P2S] tells us what it is for any X and Y to be present together. When there are multiple presents, that principle must be modified to tell us what it is for any X and Y to be present together with respect to a particular frame of reference. This yields the following principle:

[P2S$_3$] For any X and Y, if X is present in reference frame Z then Y is present in Z if and only if X is simultaneous with Y in Z.

[P2S$_3$] blocks the argument from simultaneity against the dynamic theory of time. To see this, consider again our three observers: O_1, O_2 and O_3. O_1 is present within O_1's frame of reference. So O_1 is present$_1$. Because O_2 is simultaneous with O_1 in O_1's frame of reference, by [P2S$_3$], O_2 is present$_1$ as well. But O_2 is also present$_2$, since O_2's frame of reference defines a distinct present. Importantly, O_3 is simultaneous with O_2 but not with O_1. So by [P2S$_3$], O_3 is present$_2$ but not present$_1$. Despite both being present, [P2S$_3$] does not require that O_3 and O_1 be simultaneous with one another because they are present in different senses. So the threat of contradiction is avoided.

One might worry, however, that the uniqueness of the privileged present is essential to the dynamic theory. If so, then multiplying the present is too significant a departure from standard dynamism. But even if the proposal is not a version of the standard dynamic theory strictly speaking, the important question is whether the view is similar enough to standard dynamic views to capture their core motivations. If, as discussed in Chapter 3, the dynamic theory is motivated by experience, then the current view should be able to accord with that motivation. Observers within frames of reference will experience shifts in which simultaneity class is present and thus will experience temporal passage.

That being said, there remains something worrying about the proposal. Attributing a privileged present moment to each frame of reference is to take frames of reference too seriously. A frame of reference is a coordinate system. Minkowski spacetime can be coordinatised in many distinct ways. But these coordinate systems are not really a part of reality. They amount to different perspectives on reality, which is how we have been thinking about them in the chapter thus far. It is rather like thinking that every person's spatial perspective on the world defines a unique *here*. To do so is to give too much metaphysical importance to merely perspectival notions.

To avoid this second difficulty, one might give up on the idea that each present corresponds to a simultaneity class within a frame of reference and thus seek to define the present in some other manner. Once we have started to multiply the present, however, and once we have given up the idea that the present is a simultaneity class defined by a frame of reference, there seems to be no sound basis for restricting the number of privileged present moments that exist. Indeed, a view according to which *every moment*, or every region of spacetime, defines its own privileged present would seem to become available. Such views, however, stretch the concept of the privileged present too far.

4.4.2. Diminishing the Present

Rather than multiplying the present, the next strategy disposes of it entirely. Without a privileged present, there is no way to argue from the relativity of simultaneity to the falsity of the dynamic theory of time via [P2S]. The central difficulty for any such strategy is to explain how there can be temporal passage without a privileged present; a dynamic theory of time absent a privileged present may be incoherent.

On such a view, one must try to conceptually distinguish temporal passage from the privileged present. To do so, temporal passage must be explicated in other terms. One strategy involves reducing temporal passage to certain temporal asymmetries. The details of such an account are not important for present purposes (see Chapter 5 for more on this). Rather, it is more useful to focus on the challenges that such an account faces. Such an account must show that the phenomenon to which temporal passage is reduced is capable of satisfying the motivations behind the dynamic theory of time (e.g. the reductive base must, amongst other things, underwrite the experience of passage). Whether some such phenomenon can be found remains an open question.

4.5. Relativity Revisited

So far we have looked at two broad responses to the argument from simultaneity against the dynamic theory of time. The first response was to redefine the present. The second response was to detach the dynamic theory of time from the idea that there is a single privileged present. The last response available to the dynamic theorist takes aim at sub-argument [B] offered in section 4.1; recall that this sub-argument was used to provide support for [A3], the claim that simultaneity is relative to an inertial frame of reference. The dynamic theorist has two responses to this argument available to her. She can either deny the special theory of relativity, thereby denying [B2], or she can deny the implication from the special theory of relativity to the claim that simultaneity is relative, thereby denying [B1]. It is the first option that we will focus on in what follows.

4.5.1. Neo-Lorentzian Relativity

If the dynamic theorist denies the special theory of relativity, then something must be put in its place. Some dynamic theorists maintain that there is a viable alternative: Lorentz's aether theory. Lorentz's theory preceded Einstein's special theory of relativity. Lorentz's and Einstein's theories appear to be empirically equivalent; which is to say that all of the empirical predictions made by Einstein's theory are also made by Lorentz's theory. So the two theories are, allegedly, equally capable of fitting the large amount of empirical support that we have available for Einstein's view. Lorentz's theory, however, denies the relativity of simultaneity. While Lorentz admits that observers in relative motion will make different judgements about what is simultaneous with what, he maintains that there is a fact of the matter as to who is 'correct'. Only one frame of reference provides the 'true' account of simultaneity. The trouble, however, is that which frame of reference that is, is not physically detectable.

Note that Lorentz's theory is *not* the view that there is some metaphysically privileged frame of reference, and thus some metaphysically special perspective on the world. Rather, Lorentz's view is that there are physical differences between frames of reference, and thus that, for example, there is a physical difference between Sara and Suzy's perspectives in the train case. Lorentz's view was that changes in motion alter our measuring apparatuses and clocks in such a manner that there is no discernible physical difference between frames of reference. In other words, there *is* a physical difference; it is just that any attempt to measure the difference is stymied.

Einstein's theory bested Lorentz's because it is simpler. A proponent of the dynamic theory might claim, however, that Einstein's victory was premature and that the reasons for believing the dynamic theory of time ultimately outweigh the relative benefit in simplicity that Einstein's theory affords. Science, then, has made a mistake: it is Lorentz's theory that should have won the day.

The central importance of Einstein's special relativity, however, is that it leads to general relativity, and thus to an account of gravitational fields.

Lorentz's theory was never generalised into a theory of gravity, and it is quite difficult to see how such a generalisation might proceed. The replacement of special relativity with Lorentz's aether theory, then, is not merely a matter of replacing one empirically adequate theory with another. If the replacement is made then we lose an account of gravity, unless the Lorentzian picture can be extended to handle non-inertial frames of reference (i.e. acceleration and free-fall).

Even if a general relativistic version of Lorentz's theory is produced, two problems remain. First, the present proposal recommends a revision to science on philosophical grounds. For a certain kind of philosopher – called a 'naturalist' – any such recommendation should be treated with suspicion. Naturalists maintain, roughly, that science should direct philosophy and not vice versa. Weaker forms of naturalism permit philosophy to alter science in some cases. But even a relatively weak naturalist will have reason to pause before recommending the wholesale rejection of a well-established scientific theory. Of course, the extent to which it is permissible to reject a scientific theory for philosophical reasons depends on the epistemic status of those reasons. This brings us to the second problem: it is far from clear that the reasons for believing the dynamic theory of time can overthrow an established scientific theory. As noted, the dynamic theorist maintains that we should believe the dynamic theory of time because it marries with experience. But it remains unclear just how much evidential support our experiences provide for any particular theory of time.

4.5.2. From Special to General Relativity

The strategy just discussed for denying the special theory of relativity and thus denying sub-argument [B] looks to the past before Einstein. The next strategy looks forward. Following this forward-looking strategy, the dynamic theorist notes, first, that the special relativity is false: it is superseded by general relativity. So, maintains the dynamic theorist, we should dispose of sub-argument [B] in favour of:

Sub-Argument [C]
[C1] If the general theory of relativity is true, then simultaneity is relative to an inertial frame of reference.
[C2] The general theory of relativity is true.
Therefore,
[C] Simultaneity is relative to an inertial frame of reference.

The dynamic theorist may then attempt to deny [C1], arguing that general relativity is a more hospitable environment for dynamic time than special relativity. She may do this by appealing to cosmological considerations. On some cosmological models of general relativity, the distribution of matter across spacetime is homogeneous. These appear to be physically special models of the universe. Accordingly, one might argue that such models may be used to identify a preferred frame of reference and, with it, justify the view that a particular frame provides the correct account of simultaneity.

While these cosmological models of spacetime are interesting, it is unclear that spatial homogeneity can carry the metaphysical load in supporting dynamic time. Spatial homogeneity and privileged presentness don't have much to do with each other. Moreover, even if there are frames of reference in which matter is distributed in a homogeneous manner, this type of physical privilege isn't the right kind of physical privilege to justify giving up the relativity of simultaneity. There is no reason to suppose that the laws of nature behave any differently in models of the universe where matter is distributed homogeneously through spacetime.

Suppose, however, that we do combine general relativity with the idea that simultaneity is not relative. This would be a different theory to canonical general relativity. Call such a theory: general relativity*. The choice between general relativity and general relativity* is much like the choice between Einstein's special relativity and Lorentz's aether theory. If the dynamic theorist wants to defend general relativity*, then (once again) she must recommend a significant change to our best science on philosophical grounds. But, as before, it is unclear that the evidential status of the dynamic theory can justify the change.

4.6. Quantum Gravity

So it would seem that sub-argument [C] is just as dangerous for the dynamic theory of time as sub-argument [B]. The dynamic theorist has one last trick up her sleeve. This last trick is a bit of a hail-Mary pass. One of the central goals of much of twentieth-century physics was to provide quantum theories of physical phenomena. One way to provide such a theory is to take an existing physical theory and 'quantise' it, which means (roughly) converting that theory into one that is compatible with the rest of quantum mechanics.

This quantisation project has been quite successful. A number of such quantum theories have been produced. The trouble, however, is that one cannot quantise general relativity using canonical quantisation techniques from quantum mechanics. The failure to successfully quantise general relativity has resulted in the search for a quantum theory of gravity. A number of such theories have been offered but to date there is little consensus on what the correct theory might be. The upshot of this is that we lack a quantum theory of gravity. This is, very roughly, a theory of gravity that works for very small scales, the scales at which quantum phenomena begin to arise. This has led to two distinct research programmes in physics. One project continues with the quantisation method and tries to solve the difficulties that arise from applying this to gravity. The other programme gives up on quantisation and, instead, tries to develop a theory of gravity that starts from quantum field theory and the standard model of particle physics (string theory is one such theory).

In light of the conflict between general relativity and quantum mechanics, the dynamic theorist has a modest response to sub-argument [C] available. The modest response is to note that general relativity is false: it will be superseded by a quantum theory of gravity. The implications of a false theory, however, cannot provide a knock-down argument against the dynamic theory of time. The dynamic theorist therefore cautions a wait-and-see attitude: let us see what the next revolution in physics brings and then reassess the prospects for the dynamic theory of time. In the meantime it is permissible to conduct research into the dynamic theory, so long as such research continues to be sensitive to ongoing developments in physics.

According to some dynamic theorists, if we look at certain theories of quantum gravity, we should be relatively sanguine about the future of dynamic time. Others have argued, however, that the theories of quantum gravity that look hospitable to the idea that time passes are partial theories and so it is unclear that those theories provide much by way of support for dynamism. It is safe to say, then, that the jury is very much out on whether there is any hope for a dynamic theory of time in the troubled field of quantum gravity. Part of the problem here is that most theories of quantum gravity make predictions at scales that we cannot currently test, due to technological limitations. Accordingly, even if we could find a theory of quantum gravity that somehow supported dynamic time (see the discussion in the next section) it is not clear that we could have such evidence for that theory any time soon.

It may therefore seem that the modest response to sub-argument [C] is no good. But there is hope yet. In truth, we have *no idea* what a completed theory of quantum gravity will look like. And while theories of quantum gravity arising that are friendly to dynamic time may not be very plausible, there are new theories of quantum gravity that may provide a better home for a dynamic theory of time. There are two such theories worth mentioning. First: causal set theory. According to its proponents, causal set theory provides a vindication of the concept of temporal passage within the framework of quantum gravitation. Whether this is correct or not remains to be seen. For one thing, causal set theory does not seem to involve spacetime at the most fundamental level of description, and so it is not clear that such a theory is a good home for time at all. Second: Hořava-Lifshitz gravity. Hořava-Lifshitz gravity is a theory of quantum gravity that retains the structure of general relativity whilst also reintroducing a preferred foliation. General relativity with a preferred foliation is (roughly) general relativity without the relativity of simultaneity. Such a theory, then, would at least provide a framework in which a dynamic theory of time manifests no contradictions. Whether such a theory is ultimately sustainable, and whether it can be empirically supported, is very much an open question in physics and philosophy.

4.7. Timelessness

Ultimately, however, quantum gravity may be as inhospitable to static time as it is to dynamic time. Indeed, time itself may be under threat. That's because some theories of quantum gravity appear to do away with time entirely, maintaining that time does not exist. In this last section we will briefly consider this radical proposal, and its implications for the philosophy of time.

In the history of philosophy, there have been several arguments offered that aim to demonstrate that time does not exist. Such arguments typically rely on a particularly strong conception of time, according to which the existence of the A-series is essential for time. These arguments then proceed by suggesting that either actually, or necessarily, there is no A-series and thus there is no time. These arguments are unpersuasive because, it is commonly thought, time can exist even if there is no A-series. As long as there is an asymmetric ordering of events in terms of the relations of *earlier-than*, *later-than* and *simultaneous-with* – that is, a B-series – then there is time. Strip a world of a B-series, however, and, according to B-theorists, you strip that world of time. Of course, as we saw in Chapter 1, and discuss further in Chapter 5, C-theorists think that you can strip a world of a B-series and still have a world that contains time, so long as that world contains a C-series. That is because C-theorists think that to have time only requires that there be a certain ordering of events; they do not think it is required that those events are both ordered and have a direction. B-theorists, by contrast, think that time requires not only an ordering of events but also a direction to those events: it requires that there is an ordering from earlier than to later than. As we will see, however, little hangs on the debate between B-theorists and C-theorists as to whether or not directionality (i.e. a B-series) is essential to time. For the models we will discuss in what follows are mostly ones according to which there is not even a C-series ordering of events. By everyone's lights, then, these will be worlds that lack time.

Until recently, the idea that time does not exist would have been dismissed as a philosophical fantasy. There has, however, been a resurgence of interest in this idea in light of the so-called problem of time in quantum gravity. The problem can be stated, roughly, as follows. As was noted above, our two most well-confirmed empirical theories – quantum mechanics and general relativity – are inconsistent with one another. One of the central ways in which quantum mechanics and general relativity are at odds, however, concerns their differing notions of time: quantum mechanics makes use of an absolute time variable; relativistic mechanics does not. The problem of time in quantum gravity is, roughly, how to deal with this incompatibility. To handle this inconsistency a number of physicists have argued that a completed theory of quantum gravity will lack a one-dimensional substructure of ordered temporal instances, a substructure that provides a metric for the meaningful measure of the distance between any two times. Since some kind of metric structure of this kind is necessary for both the B-series *and* the C-series, physicists appear to be suggesting that a completed theory of quantum gravity will be genuinely timeless.

What does it mean to say that according to these theories there is not even a C-series ordering of events? We will try to abstract away from the details of these theories as much as possible in what follows. But here, very roughly, is a way you might think about these theories. Imagine the whole of our universe at a time. Let's call that a three-dimensional object (it has three spatial dimensions, but no temporal dimension). That object specifies the location of every particle in these three spatial dimensions. Now, imagine all of the different ways that particles can be arranged, at a time. Suppose that there exists a three-dimensional object that corresponds to each of these different ways of arranging particles. If you think that there is time, then you think that only some of these three-dimensional objects exist, and that they are ordered by C-relations (and perhaps also B-relations). So there is some fact of the matter as to the correct ordering of those three-dimensional objects. According to timeless views, however, *all* of these three-dimensional objects exist, and none of them stand in C-relations to one another. So there is just no fact of the matter regarding how these objects are ordered. That is what it is for time to fail to exist, at least according to these kinds of views.

If time does not exist, as some philosophers and physicists believe, then that is likely to have profound implications for just about every aspect of human endeavour. Indeed, the implications are so radical that the very theories that eliminate time appear to be self-undermining. The trouble comes this way: observation seems to be an essentially causal notion. A person counts as observing that an object is F only if that object (and its F-ness) are part of the cause of that person's related sensory experience. Causation, however, seems to be an essentially temporal notion. It involves an event at one time producing an event at another time. Without time, then, causation cannot exist, and without causation, observation cannot exist and nor can observational evidence. How, then, can physicists and metaphysicians claim to have such evidence for their theories? If the notion of 'time' that is being eliminated just is the everyday notion of time, then it is very difficult to make any sense at all of the idea that empirical observation provides support for timelessness. Call this 'the observation problem'.

In order to address the observation problem, a range of phenomena must be rendered compatible with timelessness. At a minimum, it must be shown that causal processes may exist without time, along with consciousness, observation and evidence more generally. To determine the scope of the observation problem, we must first determine the extent to which timeless physical and metaphysical theories genuinely eliminate time. This highlights the importance of developing a sound conceptual understanding of temporal notions.

There are broadly two approaches to the observation problem available: the causal approach and the non-causal approach. According to the causal approach, showing that it is possible to solve the observation problem is a matter of showing that it is possible for there to be causation without time. (Recall that observation appears to be incompatible with timelessness because observation is essentially causal, and causation cannot exist without time.) According to the non-causal approach, solving the observation problem requires moving away from causal notions of evidence entirely, towards non-causal accounts of evidence and explanation. A demonstration that causation is possible without time would provide the foundations for a non-temporal theory of causation. It may, however, not be possible to provide such a demonstration. Causation and time may be inextricably linked.

There is a lot of work to be done in understanding timeless theories of physics and their implications for the metaphysics and philosophy of time. Work in this area is really in its infancy. Indeed, whatever the correct theory of quantum gravity turns out to be, we can be confident that it will spark a revolution in our understanding of the temporal structure of our universe (or lack thereof). The field of quantum gravity therefore presents an exciting opportunity for philosophers and physicists to work together. Understanding gravity is as much a conceptual problem as it is a mathematical or empirical problem. We can expect the solution, then, to come out of a renewed effort to bring physics and philosophy into close conversation.

4.8. Summary

This chapter has focused on the relationship between the philosophy and physics of time. We have looked in detail at the central argument against the dynamic theory of time from the special theory of relativity. We have also looked briefly at what the future of physics might bring for work on time within philosophy. The central issues covered can be summarised as follows:

(1) The dynamic theory of time conflicts with the physics of time via the relativity of simultaneity plus the idea that presentness is conceptually tied to simultaneity.

(2) The dynamic theorist has three options for addressing the problem posed by the relativity of simultaneity: (i) redefine the present; (ii) give up on the idea that there is *one* privileged present; (iii) deny the relativity of simultaneity.

(3) The special theory of relativity is superseded by the general theory of relativity. The general theory of relativity can be used to formulate the same argument against the dynamic theory of time.

(4) The general theory of relativity will be superseded by a quantum theory of gravity, and there is some hope for the dynamic theorist that a quantum theory of gravity will be more hospitable to dynamic conceptions of time.

(5) Work on quantum gravity in physics has led some physicists and some philosophers to conclude that time does not exist.

(6) The idea that time does not exist is a radical proposal and a great deal of work would need to be done to show that timeless theories are not self-defeating.

4.9. Exercises

i. Draw a diagram of the train thought experiment used to demonstrate the relativity of simultaneity involving three observers, instead of two.

ii. Take each of the geometric solutions to the argument from simultaneity and convert them into versions of presentism.

iii. Discuss whether we can and should give up an established scientific theory for philosophical reasons. What dangers can you see in doing so?

iv. Try to develop a version of the dynamic theory of time that makes no use of a privileged present and is thus compatible with relativity.

v. Consider the fragmentalist approach to multiplying the present. Do you think it is plausible to suppose that the universe is divided into a number of distinct realities? Why or why not?

vi. Consider the view that time does not exist. Do you find this to be a plausible view? Why or why not?

4.10. Glossary of Terms

Frame of Reference
A way of applying spatial and temporal coordinates to objects in the universe.

Inertial Frame of Reference
A way of applying spatial and temporal coordinates to objects in the universe while in constant motion.

Lorentz's Aether Theory
A physics of motion that preceded Einstein's relativistic theory.

Minkowski Spacetime
A geometric structure that makes use of three spatial dimensions and one temporal dimension, weaving them into a unique metric.

Perspective
A view-point on the universe.

Quantum Gravity
An account of gravitation at the quantum scale.

Simultaneity
Two events occurring at the same time.

Spacetime Point
A location in a four-dimensional Minkowskian geometry.

The Speed of Light
The speed of light in a vacuum, which is 300,000 km/s.

4.11. Further Readings

C. Bourne (2006) *A Future for Presentism* (Oxford University Press). Chapter 5, 'Physics for Philosophers', provides a very accessible introduction to special relativity. Chapter 6 is for more advanced students and looks at the relationship between the dynamic theories of time and relativity.

S. Savitt (2017) 'Being and Becoming in Modern Physics', *The Stanford Encyclopedia of Philosophy*, ed. Edward N. Zalta, https://plato.stanford. edu/archives/fall2017/entries/spacetime-bebecome. Section Three of this encyclopaedia entry presents a reasonably accessible introduction to the relationship between special relativity and the dynamic theories of time.

H. Putnam (1967) 'Time and Physical Geometry', *The Journal of Philosophy* 64: 240–7. For the advanced student, this is one of the earliest papers in which the conflict between the special theory of relativity and the dynamic theories of time is outlined.

L. Rudder-Baker (1974) 'Temporal Becoming: The Argument from Physics', *Philosophical Forum* 6: 218–36. Though not an introductory paper, this is a fairly accessible article on the problem posed for dynamic theories of time by special relativity.

5

Temporal Asymmetries

Chapter 4 took us on a tour of the physics of time, and how it relates to the philosophy of time. In particular, we looked at the central argument against the dynamic theory of time based on the special theory of relativity. In this chapter, we will focus on the static theory of time, and, in some sense, on its relation to physics. The goal of this chapter is to look in detail at the asymmetry of time (more on what this is, shortly). The asymmetry of time poses something of a challenge for static theories of time. Indeed, one argument for the dynamic theory of time is that since time is, obviously, asymmetric, and as static theories cannot account for this asymmetry, we ought to posit temporal passage to explain that asymmetry. If the static theorist cannot account for the asymmetry of time, or, alternatively, explain why time looks asymmetric even though it is not, then the dynamic theorist has a point: we do have some reason to posit temporal flow. Our goal here is to explore this challenge, as well as to map the space of options available for thinking about the asymmetry of time itself. We begin by getting a bit clearer on what it means to say that time is asymmetric.

5.1. Asymmetry, Anisotropy and Direction

One of the striking features of our universe is that it appears to display interesting asymmetries, many to do with time. The past seems different in various respects to the future. We know things about the past but not about the future. We can travel forwards in time quite easily but not backwards and so on. These asymmetries give rise to interesting explanatory questions that philosophers and physicists have each sought to answer. Why is the future different from the past? Why can we remember the past and not the future? Why is travel in one temporal direction so easy and travel in the other temporal direction so hard?

In order to better understand the philosophical and physical questions pertaining to temporal asymmetries, we need to differentiate between three concepts: temporal asymmetry, temporal anisotropy and temporal direction. Here is how we are going to understand the difference between these three notions. We will suppose that to talk of temporal asymmetries is to talk of processes, or phenomena, that are *temporally asymmetric*. So temporal asymmetries are asymmetries associated with things in time. More particularly, a process or phenomenon is temporally asymmetric if the behaviour of that process or phenomenon is different along one direction on the temporal axis than along the other direction on that axis.

So, for instance, suppose the process of cooking eggs is irreversible: once an egg is cooked, it stays cooked (or becomes more cooked) but cannot go back to being raw. (This is just an example; egg cooking is not irreversible in this manner, though have no fear, the chances of your cooked egg reverting to being raw are astonishingly low.) If egg cooking were irreversible in this way, then it would be a temporally asymmetric process. In one direction along the temporal axis we would always find eggs cooking, but never becoming raw, and in the opposite direction along the temporal axis we would always find eggs becoming raw, and never find eggs cooking. Even if egg cooking is reversible, it might still be that egg cooking is temporally asymmetric if we almost always see eggs cooking in one temporal direction, and almost never becoming raw in the same temporal direction. That, in fact, is how things are. In general it is sufficient for a process to be temporally asymmetric that the process is irreversible, but it is not necessary: a process can be temporally asymmetric just by its typically taking place in one, but not the other, direction along the temporal axis. As we will use the phrase, to talk of temporal asymmetries is to talk about asymmetries of process *in time*, not asymmetries *of time* itself.

By contrast, a dimension (such as time) is *anisotropic* if it has different properties along one direction of the dimension as compared to the opposite direction of the dimension. So time is anisotropic if it has different properties along the temporal dimension from past to future, than it does along the temporal dimension from future to past. At this point we should draw an important distinction between what we might call the *intrinsic* anisotropy and the *extrinsic* anisotropy of a dimension. Suppose you see a red wine glass. It might be that the glass itself is transparent, and it is filled with red wine. Or it might be that the glass itself is red. In the first case we might say that the glass is *extrinsically* red: it is red in virtue of bearing some relation to red wine (containing it). Further, we might say that the glass's redness *is accounted for*, or *reduces to* the fact that the glass contains a red substance. In the second case the glass is *intrinsically* red. The glass is red not because it bears a relation to something else (wine) but because the glass is, in itself, red.

Now suppose that instead of containing red wine the transparent glass contains a liquid that changes colour from the bottom of the glass to the top of the glass: it goes from yellow at the bottom, through green, and then blue to purple at the top. Then the colour of the glass is up/down asymmetric. We can see that by noting that the experience of an insect swimming from the bottom of the glass to the top is different from the experience of an insect swimming from the top of the glass to the bottom: the way the colour changes varies between the two directions. So the up/down dimension is colour anisotropic.

In the case in which the up/down dimension is colour asymmetric in virtue of the glass containing differently coloured fluids, we can say that the up/down dimension is *extrinsically* anisotropic. In the case in which the up/down dimension is colour asymmetric in virtue of the glass *itself* being differently coloured at different locations, we can say that the up/down dimension (of the glass) is *intrinsically* anisotropic. In general, a dimension is extrinsically anisotropic if the anisotropy is the result of the way things are distributed along that dimension, and is intrinsically anisotropic if it is anisotropic in virtue of intrinsic features of the dimension itself.

Talk of temporal asymmetries and temporal anisotropy are intimately connected. One way for time to be extrinsically anisotropic is for there to be temporally asymmetric phenomena. If lots of things in time exhibit some sort of temporal asymmetry, then the temporal dimension will be anisotropic in virtue of the asymmetries that things in time display. That leaves it open whether or not time is also intrinsically anisotropic. Perhaps it isn't. In which case it might be that time's anisotropy is fully explained by the existence of temporally asymmetric phenomena in time (though it remains to be explained why these phenomena are temporally asymmetric in this way). Or perhaps it is. Perhaps time's being intrinsically anisotropic is what explains why things in time are temporally asymmetric.

The exact nature of the connection between temporal asymmetries and temporal anisotropy is up for debate. Plausibly, the presence of the right sorts of temporal asymmetries is both necessary and sufficient for the presence of extrinsic temporal anisotropy. Equally, the presence of temporal asymmetries is clearly not *sufficient* for the existence of intrinsic temporal anisotropy, and it is unclear whether such asymmetries are even necessary. Perhaps the temporal dimension could be *intrinsically* anisotropic without there being anything *in* time that exhibits temporal asymmetry (though it is unclear on what basis one would attribute such anisotropy to time in that case, since one could not obviously gather evidence of any kind of asymmetry in time). At any rate, those who think that time is intrinsically anisotropic do, in fact, also tend to think it contains temporally asymmetric phenomena, so to a large extent we can avoid taking a stand on this issue.

This brings us to the third concept that we promised to differentiate: directionality. We will say that time has a direction just in case time is aniso-tropic (either intrinsically or extrinsically) and there is some fact of the matter (given by the world) as to which direction is which. So, for instance, think about walking up, or down, a mountain. The path up the mountain is anisotropic: the properties going up the mountain are different from those going down the mountain. Does the path have a direction? It does if there is an *objective* fact of the matter regarding whether the path goes from the bottom of the mountain to the top, or goes from the top to the bottom (which there isn't, unless the path is impassable going up versus going down).

Or return to consider our coloured glass. An insect can swim from the bottom to the top, or from the top to the bottom. The up/down dimension of the glass will have a direction, as well as an anisotropy, if there is a fact of the matter whether the dimension *really* goes from up to down, or down to up. In a way, then, the dimension is directed just in case, metaphorically speaking, in addition to it being anisotropic there are little 'arrows' that point from the top to the bottom, or from the bottom to the top so that we know whether the glass 'goes' from yellow to purple, or from purple to yellow. In both cases it is hard to see what that fact might amount to. To return to the mountain path example, one might have reason to walk from the top to the bottom (to limit effort) but there is no sense in which the path *itself* really goes from the top to the bottom (or the bottom to the top). So although there is anisotropy, there is no directionality. By parity, time has a direction if it is anisotropic, and there is a fact of the matter as to whether time is directed from the past to the future, rather than from the future to the past. If that is right, then although it is *necessary* to time's having a direction that it be anisotropic, it is not *sufficient*.

In what follows we begin by considering various temporal asymmetries: that is, various phenomena or processes that are temporally asymmetric. We will then move on to consider ways in which we might explain time's direction (if indeed time has direction).

5.2. Temporally Asymmetric Phenomena

Our world seems to contain many temporally asymmetric phenomena. That is, it seems to contain phenomena that behave differently along one direction on the temporal axis as compared to how they behave along the other direction on the temporal axis. Examples are everywhere, and mostly we probably don't even think about them. But consider the following. We age along the temporal dimension in one temporal direction. All of us get older towards the direction we call future, and away from the direction we call past. What's true for each of us is also true for just about everything else, from milk souring, to eggs going off, to corpses rotting. These are all specific instances of a more general phenomenon known to us all through high-school physics as the second law of thermodynamics, which says (roughly) that disorder (i.e. entropy) increases towards the future.

There are plenty of other examples of temporally asymmetric phenomena. Waves spread *from* an emitter, but not *towards* an absorber. When any of us throws a rock into a still pool we see waves emitted from the point of contact with the water. We would all be surprised to see waves converging on a point, and a rock being disgorged from the water.

There are also cosmological asymmetries. It seems that the universe is expanding, and that it is expanding *away* from the past and *towards* the future. So cosmological expansion seems to be temporally asymmetric.

Causation, too, seems to be temporally asymmetric. We expect causes to precede their effects in time. Even if backwards causation is possible, and, indeed, even if there is actually some backwards causation, *in general* causation seems to be temporally asymmetric. Related to this asymmetry of causation is the asymmetry of determination, or the fork asymmetry. This asymmetry is captured by the idea that a single cause can have myriad effects, but it is very rare indeed for a single effect to have myriad independent (full) causes.

For instance, imagine you spill a plate of spaghetti bolognaise all over the floor. One way of cleaning it up is to crawl all over the floor and pick up each bit of spaghetti and tomato sauce. Since the bits of spaghetti have spread far and wide, this will be quite a big job. It would have been much easier simply to prevent the plate from hitting the ground in the first place and spreading its contents. The lesson here is quite general: it's much harder to mop up after some cause has left multiple effects than it is to prevent a single cause from obtaining in the first place (witness oil spills to fully appreciate this). Consider, once more, the floor covered in spaghetti and sauce. One way the floor could have ended up like this is if many individuals had each, independently (i.e. not in conspiracy with one another, or in virtue of some common cause), come into the kitchen and thrown a piece of spaghetti on the floor. Then it would actually be easier to clean up the floor by cleaning up all the spaghetti than it would be to try and prevent each and every person from throwing a single piece of spaghetti onto the floor. The point, however, is that we rarely see cases in which there are multiple independent causes and a single effect, while we frequently see cases in which there are multiple effects of a single cause. In this respect the causal structure is temporally asymmetric.

Finally, there are many temporally asymmetric psychological phenomena. For instance, we deliberate towards the future and not towards the past. That is, we deliberate about events that are future relative to our location in time, and not about events that are past relative to our location in time. We also seem to differently value past and future events. Suppose you know you have to undergo a painful dental procedure. It is common to prefer that that procedure be located in the past, rather than the future. Or consider a very tasty chocolate cake. It is common to prefer that the event of eating the cake is in the future, rather than the past, and indeed in the near future rather than the far future. It is often thought that if we had to choose between having had a very painful dental procedure in the past, and having to have a less painful dental procedure in the future, we will choose the former over the latter, even though overall we are subject to more pain that way. (We also typically value things that lie in the further future less than we value things that lie in the near future. This is known as discounting. This, however, will be of less interest to us here, since it doesn't really reveal an asymmetry between the two temporal directions.)

To this list, dynamic theorists of time would also add the asymmetry of temporal flow. As time passes, events that were future become present and then past. We do not see the reverse happening. That is, as time passes we do not see events that were past become present and then recede into the yawning future. Of course, whether or not time really flows is, as we have seen, a very controversial matter. If it does, however, then we have one more asymmetry to deal with. In this case, though, we'd be looking at an intrinsic anisotropy of time itself, one that has a preferred direction.

In what follows we focus in more detail on four of these asymmetries: entropy and the second law of thermodynamics; the asymmetry of causation; the fork asymmetry; and the asymmetry of deliberation. We will briefly return to the asymmetry implicated in the flow of time in section 5.4.

5.2.1. *Entropy and the Second Law of Thermodynamics*

Informally put, the second law of thermodynamics says that entropy always increases over time. (In fact, what the second law actually says is that the total entropy of an isolated system will always increase over time, unless that system is in equilibrium, in which case it will remain stable: the system will never decrease in entropy.) What does that mean? Entropy is a measure of *disorder*. The more entropy there is, the more disorder there is. So the second law of thermodynamics says that an isolated system will never decrease in disorder over time.

We can now see why ageing processes are a specific instance of increasing entropy. Think about the rotting corpse. Bodies start off in a fairly low entropy state (being alive), and gradually, as time passes, become more disordered as they fall apart and return to the soil.

If the second law of thermodynamics is a law, then the increase of entropy is one example of a temporally asymmetric phenomenon: entropy increases (or stays the same) towards the future, and thus decreases (or stays the same) towards the past. The direction towards the future is different from that towards the past when it comes to entropic processes. In section 5.6.1 we return to consider entropy and the second law of thermodynamics in much more detail. There, we revisit the question of whether the second law of thermodynamics is really a law and whether there is anything temporally asymmetric about the distribution of entropy.

5.2.2. *Deliberation*

Consider that when you get up tomorrow, you will deliberate about what to have for breakfast: toast or cornflakes. You will deliberate towards the direction you call the future. You will not, however, deliberate about whether you ought to wear a jumper to bed the night before. That's not a special feature of you. We all deliberate about what we will do at later times, that is, at times towards the future, and none of us deliberate about what we did do at earlier times, that is, at times towards the past. So we might say that there is a deliberation asymmetry: we deliberate towards the future and away from the past. That asymmetry, in turn, seems to be related to another asymmetry: an asymmetry in our knowledge. Why do we deliberate about the future but not the past? Well, one answer that immediately springs to mind is that we already know what we did in the past, and so it would be pointless to deliberate about it.

The point of deliberation is to decide what to do. But one can't try and decide what to do with respect to P if one already knows that one did P. That's why you can't deliberate about whether to wear a jumper last night: because you already know whether you wore the jumper or not. By contrast, you don't know what you will have for breakfast tomorrow (you might have strong suspicions, based on what you have had every other morning, but if you decide to treat breakfast-eating as a matter for deliberation, you will take it that you do not know what you will eat until after you have made the decision). Another closely related asymmetry, which, one might think, helps explain the deliberation asymmetry, is the causal asymmetry. Causation itself seems to be temporally asymmetric. Causal processes go from past, to future.

5.2.3. Causation

Causation typically goes from past to future. As we will consider in Chapter 8, where we look at time travel, it might be that sometimes causal processes will go in the opposite direction – from future to past. Nevertheless, in general, causes temporally precede their effects, rather than the other way around.

Indeed, that causation typically (or always) goes in one temporal direction might be thought to explain the deliberative asymmetry. After all, the reason why you deliberate about something is that you suppose yourself to have some causal control over whether or not to bring that thing about. You don't deliberate about whether to cool the sun down because that's not something you take yourself to have any control over. But if you only deliberate about things you can causally affect, and if causes typically precede their effects in time, then it makes sense that you only deliberate about future events and not past events. For if causes typically do not go backwards, then you will typically have no control over events in the past. You can't deliberate about whether or not to wear a jumper last night, because you have no causal control over your wearing of a jumper last night.

5.2.4. Forks

The temporal asymmetry of causation is closely related to the fork asymmetry. Think about a fork. It has a single handle, and then a bunch of prongs at the end. (If you own a fork with more than three prongs, that fork is ideal for our present purposes.) The fork is anisotropic in that from one direction it runs from a single handle to a number of prongs, and from the other end it runs from a number of prongs to a single handle. Now imagine that the handle of the fork is pointing back towards the past, and the prongs towards the future. The thought is this: what we find in our world is that very many later events (the prongs on the fork) are all correlated with a single earlier event (the handle of the fork). So, for instance, imagine that you set off a firework at a crowded market. The letting off of the firework is a single event, but now if we watch to see what happens we will see many effects emanating from that cause. People in the market place will run in many directions, grabbing different possessions, and engaging in different behaviours.

That is the fork asymmetry: from a single cause, we see many dissipating effects. It's an asymmetry because we don't see *reverse forks*. In the case of the firework we see a number of correlated events (the actions of the people in the market place) preceded by a common cause (the letting off of the firework). But we don't see cases in which there are correlated events, linked by a joint effect. That would be a case in which lots of independent, but correlated, events all come together to produce a single effect. We can imagine what that would be like by just imagining our market place scenario in temporal reverse: from people being scattered we see those people gradually converge on a market place, and come ever closer, until there is an implosion of fireworks into a box of undisturbed skyrockets. We would be very surprised to see events play out this way. What could bring about such an inverse fork? Imagine that philosophers had lots of money, and hired a film crew to depict an inverse fork, not by running film backwards, but by filming the market place just described by having actors running backwards towards what will be the firework going off. Notice that this still wouldn't be an example of a reverse fork, since the correlated events would not, in fact, be independent: they would be the effect of a common cause, to wit, the philosophical director who is attempting to create a reverse fork. Notice that in the absence of such a director we would be surprised indeed to see the market place occurring 'in reverse'. That's because we don't typically see reverse forks, while we do often see ordinary forks.

5.3. Time's Direction

That there are so many temporally asymmetric phenomena seems to demand an explanation. One potential explanation, and one that seems to accord with the way the world seems to us, is that time has a direction. If time itself has a direction, then this can explain why things in time are arranged in a temporally asymmetric manner. The alternative view is that time has no direction, but some features of our local environment, or our psychologies, or the combination of both, makes it appear as though time has a direction. As we saw in Chapter 1, the view that time has a direction is shared by both A- and B-theorists. A- and B-theorists agree that time has a direction, but disagree about what accounts for time having a direction. By contrast C-theorists hold that time has no direction, but merely appears to do so. We consider this view in section 5.3.3.

There are two very different views about temporal direction. We call the first of these views *primitivism* about direction, and the second *reductionism* about direction. One version of primitivism about direction is the view that time has a direction because it has temporal flow, and its having temporal flow is a primitive matter. This version of primitivism is the view that A-theorists endorse, and we consider it in section 5.4. A second version of primitivism is a version defended by certain B-theorists. We discuss this B-theoretic primitivist view in section 5.5. Finally, probably the most popular view about temporal direction is the view that temporal direction reduces to something else. This is a static B-theoretic view according to which time has a direction, but its having that direction is accounted for by some other phenomenon or phenomena. In the following sections we will outline some primitivist and reductionist proposals. First, though, we need to get clear on the difference between primitivist and reductionist views.

5.3.1. Primitivism and Reductionism

Primitivism is the view that time has a direction, and that its having that direction is intrinsic to time itself. Reductionism is the view that time has a direction, but its having that direction is reducible to something else.

We have already met the difference between intrinsic and extrinsic anisotropy. The difference between primitivism and reductionism about direction is very similar. Recall that those who hold that time has a direction think not only that time is anisotropic, but, in addition, that there is a fact of the matter as to which direction time points. Primitivists think that time's directedness is a primitive matter. So, for instance, remember the example of the coloured glass that gradually changes from one colour at the bottom to another colour at the top. The primitivist about up/down directionality will say that there is some primitive matter of fact that the up/down dimension really goes *from* up *to* down (or vice versa). There are, as it were, little arrows embedded in the up/down dimension which point from up to down (or vice versa), and these arrows cannot be reduced to or explained in terms of anything else. An obvious primitivist view is to pair the view that temporal *anisotropy* is extrinsic with the claim that temporal *direction* is primitive. So what makes the two directions along the temporal dimension different is some feature of how things are distributed along that dimension, but what makes one direction the way that time in fact goes is a primitive matter.

By contrast, reductionists about temporal direction hold that time has a direction, and its having that direction is reducible to some feature of things in time. It is easy to see why reductionism about anisotropy is appealing, and easily motivated. For the reductionist can simply say that the difference between the two directions along the temporal dimension is the result of processes being temporally asymmetric. It is less easy to see how the reductionist can make sense of directionality. Even if it's true that processes 'age' towards what we call the future, and away from what we call the past, it in no way follows that time goes from what we call the past, to what we call the future, rather than the other way around. It seems conceptually possible that time might go from the past to the future, while beings Benjamin Button their way through time: ageing towards the past and growing younger towards the future. One possibility that suggests itself to the reductionist is to appeal to the laws of nature. Should it turn out that the laws of nature are themselves temporally asymmetric, the reductionist could argue that this is what grounds time's having a direction, rather than merely being anisotropic. It is the laws that provide the needed 'arrow'. We consider this suggestion in the following section.

5.3.2. Asymmetry and the Laws

The idea that the reductionist can appeal to asymmetric laws to ground the direction of time is appealing. But there's a problem. The problem is that the laws of nature are *time-reversal invariant*. What that means is that if we take any physical law described as an equation or formula, which includes a temporal variable (t) in that equation or formula, and if we then replace t with –t in the formula or equation, the resulting equation still describes a law. What does this mean? It means that for any law-like physical process, such as, for instance, dropping an egg on the ground and the egg breaking, we can describe the reverse of this process – a broken egg leaping up from the ground and reforming into a whole egg – and that reversed process also conforms to the physical laws. To put it another way, the laws apply equally well regardless of which direction we take to be future, and which we take to be the past, in describing the evolution of some process.

To put this in context it's worth noting that the laws are also space-reversal invariant. That means that if we reverse any of the spatial dimensions in the formulae expressing the laws, the laws remain unchanged. No doubt you will not find that surprising. Quite the reverse; presumably you would find it surprising if the physical laws were different going along one direction of a spatial dimension as opposed to the other direction along that dimension (imagine the laws are different going from up to down, as opposed to from down to up). The same is true for the temporal dimension as for all three spatial dimensions. What time-reversal invariance means, then, is that a mirror image of our universe in which all objects have their positions and momenta reversed would evolve under the same physical laws. That is why the world we described in Chapter 3, in which physical processes in one temporal half of the world occur in reverse order to those in the other temporal half, is taken to be nomologically possible. The laws of nature in such a reversed world *may be the laws we have in our world*.

At this point it is worth noting that it's not *quite* right to say that the laws are time-reversal invariant. There is one very small exception: the laws featuring the kaon (a particular type of sub-atomic particle) violate time-reversal invariance. Some take this to be an important discovery, and hang their hat on the idea that what time's direction consists in can be understood in terms of this time-reversal invariance. But this seems a bit of a stretch. Even if the kaon does behave in a non-time-symmetric manner, it's hard to see how that could be what gives time a direction. Amongst other things, it's hard to see why this fact about the humble kaon would explain why we typically see corpses rotting, rather than coming back to life, and why we typically see waves diverging from an emitter and not converging on an absorber, and so on. So for present purposes we will simply suppose that the laws are time-reversal invariant and set aside the perplexing features of the kaon.

The problem, then, is this. If the laws of nature are time-reversal invariant, then we cannot reduce temporal directionality to asymmetrical laws. Indeed, given the symmetry of the laws, the very presence of temporal anisotropy is puzzling. If the laws are symmetrical, it seems that we ought to predict that phenomena will be temporally symmetric, rather than temporally asymmetric. Yet we see temporally asymmetric phenomena all over the place. So now we are left with two questions. First, we need to answer the question of why there are temporally asymmetric phenomena if the laws are temporally symmetric, and, second, reductionists need to find some other reductive base to account for that direction. Indeed, if we are drawn to reductionism about direction, all of this might suggest that having failed to find a reductive base for directionality we ought to conclude that time does not, in fact, have a direction. It is to this view that we now turn.

5.3.3. Directional Eliminativism

One response to the discovery of time-reversal invariant laws is to bite the bullet and concede that in fact time doesn't have a direction after all. This is precisely what C-theorists, who we might also call *directional eliminativists*, do. C-theorists hold that events stand in temporal relations with one another – C-relations. There are genuine temporal distance relations between events, just as there are genuine spatial distance relations between places. But time is like space, in so far as it has no direction. So, for instance, just as Singapore and Sydney are separated by a certain spatial distance, so too are Caesar's death and Hillary Clinton's birth separated by a certain temporal distance. And just as there is no fact of the matter about which way space is directed (so that we can rightly say that space goes from Singapore to Sydney, or, conversely, from Sydney to Singapore), likewise there is no sense in which time is directed, so that it goes from Caesar's death to Clinton's birth (or vice versa). Of course, C-theorists think that time *appears* to have a direction, but that is all it is: mere appearance.

It is worth noting, at this point, that directional eliminativists need not deny that time is anisotropic. If there are temporally asymmetric phenomena, then time will be extrinsically anisotropic even though it lacks a direction. Indeed, the C-theorist can say that what we *call* the past is just the direction towards (for instance) decreasing entropy, and the direction we *call* the future is the direction towards increasing entropy. So the C-theorist can allow that we correctly say things like 'Sara's birth was in the past' and 'the big crunch is in the future'. What the C-theorist denies is that these claims are made true by time having a direction that runs from Sara's birth to the big crunch, and in virtue of which the big crunch is *objectively* in the future as opposed to the past. Indeed, the directional eliminativist will likely allow that in a world in which one temporal half is a mirror image of the other temporal half, true claims about which direction is past, and which future, are reversed. That is, what Sara calls 'past' will be what those at the other temporal end of the universe rightly call 'future'.

Of course, merely claiming that time has no direction does not absolve the directional eliminativist from explaining the temporal asymmetries that we see around us. Even if time has no direction, we still need to explain why it seems, to many at least, as though it does; and we need to explain why so many phenomena are, or seem to be, asymmetrically oriented in time, if time itself lacks a direction.

To do so, the directional eliminativist must offer an account of why various phenomena are, or at least seem to be, temporally asymmetric, one that does not appeal to time having a direction. We consider how this story might proceed, at least with regard to some temporally asymmetric phenomena, in section 5.6.

5.4. The Flow of Time

One possibility for explaining the direction of time that we have thus far neglected appeals to the dynamic theory of time. The thought is that time has a direction because time flows. The future just is the direction towards which time flows. The difference between the past and the future, then, is that time flows towards the future and away from the past. This proposal shares with primitivism the claim that time's having a direction is an intrinsic property of time itself. That doesn't mean that the dynamic theorist has to think that temporal flow is primitive. She might think this, and perhaps that is what some moving spotlight theorists do think. But she need not. For instance, some presentists think that temporal flow is the changing of a single three-dimensional slice. To be sure, what it is for a slice to change is itself a primitive matter, but we can reduce the flow of time itself to the changing of this thin wafer of reality. Equally, the growing block theorist might say that what it is for time to flow just is for the universe to accrete new slices. Again, she will likely say that it is a primitive matter that the universe grows in this manner, but she can nevertheless reduce time's flow to the accretion of these slices. Having done so she will go on to say that time's flow is intrinsic to time itself, and therefore that time's direction is intrinsic.

One cost to views such as this is that if time's direction is intrinsic to time itself, and so time's direction owes nothing to the asymmetry of things *in time*, then it remains a mystery as to why time's having a direction explains why there are temporally asymmetric phenomena. This mystery is particularly forceful if we accept that the laws are time-reversal invariant. Were that not so, we might try to explain why certain phenomena are temporally asymmetric by noting that time has a direction, and that it is a law of nature that phenomena behave differently along one temporal direction than the other. If one accepts that the laws are, however, symmetrical, then it is unclear just what explanatory work any primitivist view of direction can achieve.

We have already seen this problem arise in the context of temporal phenomenology, in Chapter 3. There, recall, we considered the mirror world, and asked whether the temporal phenomenology of individuals in one temporal half of the mirror world would be different from the temporal phenomenology of their physical duplicate doppelgangers in the other temporal half of the world. The worry, recall, was that if the phenomenology is not different, then it seems as though temporal flow makes no difference to the way things seem to us (i.e. to our temporal phenomenology).

We can now generalise this worry by considering the mirror world in a bit more detail. Call one direction along the temporal dimension D, and the other direction D*. Let's suppose that the direction of temporal flow is towards D, and away from D*. Then it is very hard to see how the presence of temporal flow, and hence temporal direction, can explain the presence of temporally asymmetric phenomena. After all, in one temporal half of the mirror world the temporally asymmetric phenomena 'point' towards D, and in the other half they 'point' towards D*. So those asymmetric phenomena are aligned with the direction of flow in one half of the world, but not in the other half of the world. To put it another way, in one temporal half of the universe, given the direction of time, eggs *really do* typically go from being raw, to being cooked, to being eaten. But in the other temporal half of the universe eggs *really do* go from being eaten, to being cooked, to being raw with the passage of time. The direction of the flow of time seems to be irrelevant to these processes, and thus to be incapable of explaining them.

Of course, the defender of temporal flow (and hence temporal direction) might argue that we are demanding answers to the wrong questions. She might appeal to whatever account the directional eliminativist appeals to in explaining why there are temporally asymmetric phenomena, then insist that appealing to temporal flow does *nothing more* than give time a direction. That time flows in one direction and not the other is not what explains why certain phenomena are temporally asymmetric; all it explains is why one direction is objectively future, and the other objectively past. If the dynamic theorist takes this view, however, she admits that the flow of time does not explain why phenomena in the world are temporally asymmetric. There is one less thing that the flow of time can explain. This is particularly troubling because it is precisely this kind of asymmetry that one would expect the flow of time to be implicated in. So if one detaches the flow of time from the asymmetry of things in time, it will be much less clear that the explanatory benefits of positing temporal flow outweigh its ontological costs.

5.5. Primitive Direction

Much the same problem arises if instead of appealing to temporal flow to ground temporal direction, we instead suppose time to have a primitive direction. The difference between the view that time has a direction because time flows, and the view that time does not flow but has a primitive direction, is important.

The view that time has a primitive direction, but not in virtue of time flowing, is a version of the static B-theory of time. Remember once again our coloured glass. As we have already noted, the primitivist about up/down direction thinks not only that the up/down direction is anisotropic, but that the glass is directed, say, *from* up, *to* down. What makes this the direction of the dimension is that (metaphorically speaking) there are little arrows embedded in the coloured glass pointing from up, to down. Notice that none of this requires that the liquid in the glass, or the glass itself, flows. The arrows themselves (metaphorical though they may be) are entirely static. They point in a certain direction, but they do so without moving. So the view that time has a direction is entirely consistent with the B-theory of time, as long as time's having a direction is not a matter of it having temporal flow.

But the very same worry arises for the view that direction is primitive as for the view that direction is accounted for by flow. That's because if time's direction is primitive, rather than being reducible to something else, then it seems plausible that the direction of time can entirely come apart from the various temporally asymmetric phenomena that we see. Return again to our mirror world. This time, however, rather than supposing that in such a world there is temporal flow, instead suppose that in that world time has a primitive direction. The future is towards D, and the past towards D*. So in one half of the world, the temporal asymmetries align with the direction of time. But in the other half of the world, they fail to align with the direction of time: instead, time goes in the opposite direction to the asymmetries present.

The defender of primitive directionality could, of course, insist that the direction of time cannot come apart from various temporally asymmetric phenomena in this way. She could offer a response similar to one offered by defenders of temporal flow (we met a version of this response in Chapter 3). Recall that one problem with positing temporal flow is that if we suppose the laws of nature to be time-reversal invariant, then we must countenance the nomological possibility of the mirror world, and that, in turn, raises the problem that it seems as if temporal flow is explanatorily idle. But of course, one response on behalf of the defender of temporal flow would be to argue that if there really is temporal flow in our world, then we must be wrong about the laws. If time does flow, then the laws surely cannot be time-reversal invariant. And if they are not time-reversal invariant, then we have little reason to suppose that the mirror world is nomologically possible.

Likewise, the defender of primitive directionality might argue that there are nomologically necessary connections between the primitive direction of time and various temporally asymmetric phenomena, so that these always go together (in nomologically possible worlds). They *cannot* come apart. So the mirror world is not nomologically possible and the direction of time does align with the various temporally asymmetric phenomena. Again, though, this requires that we reject the contention that the laws are time-reversal invariant. Methodologically speaking, we are making recommendations for science on philosophical grounds. If we aren't willing to break the news to physicists, then as soon as we allow that the mirror world is nomologically possible, the primitivist about direction seems to lack any explanation for the temporally asymmetric phenomena we see around us. Or, at least, she cannot appeal to time's having a direction to explain these phenomena. Bearing this in mind, in what follows we turn to consider reductionist approaches to the direction of time to see if they fare any better.

5.6. Reduction

Reductionists hold that there is a direction of time, and that time's having that direction is reducible to some temporally asymmetric phenomenon or other. Exactly what it means to say that X is reducible to Y is controversial, but, roughly speaking, if X reduces to Y, then there being X is in some sense nothing more than there being Y. For instance, some philosophers think that mental states reduce to brain states: they think that what it is to have a mental state just is to have a certain brain state. Or one might think that there being a table reduces to there being some arrangement of particles: there being that arrangement is all it takes for there to be a table. If X reduces to Y we will say that Y is X's *reductive base*. Since reductionists about temporal direction think that the reductive base is some sort of asymmetric phenomenon – they just disagree about which sort it is – we can say that different reductionists posit different *base asymmetries* as the reductive base for the direction of time. We can now ask what options there are open to the reductionist for reducing the direction of time to some base asymmetry. Before we do that, though, we should get clearer on just what it is that the reductionist is proposing. She tells us that the direction of time reduces to some base asymmetry. Since we don't know what that base asymmetry is yet, let's just call it B. What does it mean to say that the direction of time reduces to B? There are two different options the reductionist might endorse:

(1) **Identification Thesis:** B is identical with the direction of time.
(2) **Grounding Thesis:** B grounds the direction of time.

According to (1), whatever B is, B is identical with the direction of time. So if, for instance, B is increasing entropy, then the direction of time is *identical* with increasing entropy. Then time has a direction in any world in which there is increasing entropy, and lacks a direction in any world that lacks increasing entropy. By contrast with (1), (2) says that B grounds the direction of time. Grounding is a relatively new philosophical posit, introduced by metaphysicians in an attempt to devise a relation that can accommodate *dependence* between things in the world. So, for instance, we might want to say that the table depends on the particles being arranged a certain way, but not that the table is identical with those particles arranged that way. So we could say that the table is grounded by the particles. What's nice about this is that it allows us to say that the table could have depended on some other bunch of particles entirely (since it's not identical to those particles). Likewise, (2) allows us to say that although time's direction is in fact grounded in B, in other worlds time's direction is grounded by something else. For instance, it allows us to say that in our world it is increasing entropy that grounds time's direction, but in other worlds some other physical asymmetry grounds time's direction (such as, for instance, some cosmological asymmetry). It's worth bearing this difference in mind as we consider a few candidates for the base asymmetry.

5.6.1. Entropy

We earlier mentioned the second law of thermodynamics and flagged the question of whether it really is a law. We can now examine that question in more detail. If it were a law that entropy never decreases towards the future, then it might seem that the temporal asymmetry of entropy is a good candidate to be the base asymmetry for the direction of time. The reductionist might say that the direction towards the future is the direction towards which entropy increases, and the direction towards the past is the direction towards which entropy decreases. But, a problem arises. Why think that the future is the direction towards which entropy *increases*, rather than the direction towards which entropy *decreases*? That is, why think that time goes from low entropy to high entropy, and not the other way around? Nothing about entropy itself tells us that time is directed from low to high entropy. Directional eliminativists make just this objection. They point out that even if entropy is asymmetric in this way, all this tells us is that time is anisotropic, not that it has a direction. For if time has a direction then there must be some objective fact of the matter that time goes *from* low entropy *to* high entropy (rather than the reverse). But the discovery that entropy is temporally asymmetric in no way guarantees this fact.

Even setting that problem aside, however, another one arises: namely, it's not clear that it *is* a law that entropy is temporally asymmetric. According to contemporary statistical mechanics the second law of thermodynamics is not really a law at all, but merely reflects the probabilities of certain macrostates, conditional on local boundary conditions having certain properties. Let's unpack what that means. A *microstate* is a specification of the position and velocity of each particle in the system. A *macrostate* is a specification of the observable properties of the system, such as volume and temperature. Each macrostate can be produced by many different microstates. So imagine a gas that is distributed through some container (which is the system in question). The microstate is a specification of the position and velocity of each particle of gas in that container. The macrostate is a specification of the observable properties of the gas – its distribution, temperature, and so on. So, for instance, suppose the gas is spread out uniformly through the container. Then it has a uniform macrostate. But there are lots of ways we could distribute the very same particles to attain that same macrostate. Consider some particular particle, Freddie. There are lots of locations in the container at which Freddie could be located, consistent with the gas being uniformly distributed. Now let's suppose that the probability of every microstate is the same. Conditional on that being the case, the probability of any macrostate will be proportional to the number of microstates that produce that macrostate. So, if there are more microstates that produce macrostates in which the gas is evenly distributed across the whole container than there are microstates that produce macrostates in which the gas is distributed across only half of the container, then it is more probable that the gas will be distributed across the whole of the container than just across half.

Entropy, recall, is a measure of order: the more order, the lower the entropy, the more disorder, the higher the entropy. So suppose we want to know how likely it is that a state will have high entropy, rather than low entropy. Well, notice that high entropy macrostates are ones that can be produced by more microstates than low entropy macrostates. Consider a very low entropy macrostate in which the particles in the container are lined up from left to right. There are far fewer microstates that can produce this state than there are microstates that can produce a uniform distribution of the gas. That is, a smaller proportion of the total microstates can produce this very low entropy macrostate, as compared to the proportion of the total microstates that can produce the higher entropy macrostate. Given this, we should expect the lower entropy macrostate to be less probable than the higher entropy state. In general, we should expect low entropy macrostates to be less probable than high entropy macrostates, because low entropy macrostates are produced by fewer microstates.

This explains why systems tend to move towards high entropy macrostates: they evolve towards states that are the most probable states, and higher entropy macrostates are more probable because a higher proportion of the total microstates are high entropy macrostates.

This is why the second law appears to be true. Equally, it tells us why it isn't really a law at all (assuming, that is, that laws are supposed to be exceptionless generalisations). Statistical mechanics only tells us that it is more probable that a state will move towards a higher entropy state, not that it is *impossible* that it will move towards a lower entropy state. If the universe is big enough, then even though there is a very low probability of decreasing entropy, it is possible that there are parts of the universe in which entropy decreases.

Moreover, what holds for the direction into the future also holds for the direction into the past. For the very same reasoning we just used, which told us to expect future higher entropy states, should also lead us to expect that entropy will, in general, increase in *both* directions in time from an ordered state. So we should expect that entropy will increase into both the future and the past. Statistical mechanics is temporally symmetric. That's puzzling, since it's not what we observe.

So why does entropy in our world increase towards what we call the future, and decrease towards what we call the past? Why might we expect to see entropy decreasing towards the past? Well, according to statistical mechanics we would *expect* to see this if, in the past, there were a highly ordered state, and entropy has been increasing *away* from that state. So this suggests that somewhere in the past there is a very low entropy state. It only really matters that that state is somewhere in our distant (but not too distant) past; but usually advocates of this solution hold that the low entropy state is a *boundary* condition – an initial or final condition of the universe. So typically it is held that the big bang generated a very low entropy condition very close to the 'beginning' of the universe, and entropy has been increasing away from this state ever since. That there is such a low entropy condition is called the Past Hypothesis (PH). According to PH, shortly after the big bang the initial state of the universe was in a very low entropy state. Given statistical mechanics, we should expect entropy to increase away from that state, that is, towards the direction we call the future, and we should expect entropy to decrease towards the direction we call the past.

The Past Hypothesis, in conjunction with statistical mechanics, explains why we typically see entropy increasing towards the future and decreasing towards the past. This appeal to statistical mechanics is, however, excellent news for the directional eliminativist, and not good news at all for the reductionist about temporal direction. That's because we can explain the existence of temporally asymmetric entropy without supposing time, or the laws of nature, or anything else, to be asymmetric or directed. Moreover, if the second law of thermodynamics is not an exceptionless law, but a mere probabilistic generalisation, it remains unclear whether the distribution of entropy is a plausible base to which to reduce temporal direction. The PH posits the existence of a low entropy 'initial' condition. But nothing in the laws of nature prohibits there also being a low entropy 'final' condition: a big crunch. If entropy is very low at both 'ends' of the universe, then we get a world that looks like the mirror world we saw in Chapter 3. But then, which direction does time go in, in such a world? What makes one low entropy condition the 'initial' condition, and the other the 'final' condition, as opposed to the other way around? Nothing about the distribution of entropy itself seems to tells us that time is directed from one end of the universe to the other, rather than the other way around. If entropy decreases towards a big crunch at the other end of the universe then why isn't the other end of the universe an 'initial' condition, and our end a 'final' condition? What makes it the case that the direction of time runs from what we call past to what we call future, rather than the other way around? It seems most natural to say that from the perspective of one temporal half of the universe, one direction is future, and from the perspective of the other temporal half of the universe, the opposite direction is future. But at best this would give us a number of local directions to time rather than a single global direction. At worst we might be inclined to say, with the directional eliminativist, that there is no temporal direction, there are just phenomena that are, at different locations, temporally asymmetric and which make us mistakenly think that time has a direction. In either case, matters look tricky for anyone attempting to reduce time's direction to the increase of entropy.

5.6.2. Causation

We have already mentioned in this chapter that causation is temporally asymmetric: causes typically precede their effects in time. So another option for accounting for the direction of time would be to appeal to the direction of causation. Suppose, for a moment, that such a reduction is possible (we return to this shortly). Still, there is a problem. The reductionist urges that we take the direction of time to be aligned so that the future is the direction towards which there are effects, and the past is the direction towards which there are causes. Time goes, as it were, *from* cause, *to* effect. But why think that? Why not, instead, think that time goes *from* effects, *to* causes? Why not think that the future is the direction towards causes, and the past the direction towards effects? The directional eliminativist will, as always, urge that merely showing that cause and effect are temporally asymmetric might show that time is anisotropic, but not that it is directed: for we need some reason to think that time goes from cause to effect rather than the other way around, and nothing about the nature of causation itself seems to give us that reason.

But suppose the reductionist could solve that problem. Nevertheless, the reductionist project faces other issues, analogous to those we noted in the case of reducing temporal direction to the distribution of entropy over time. For just as statistical mechanics tells us that we should not expect entropy *never* to decrease towards what we call the future, so too, there are reasons to think that causal relations are not, or need not always be, aligned in the same direction. If backwards causation is nomologically possible, or indeed actual, then we should expect worlds like ours to be ones in which effects sometimes precede their causes. But suppose that there are large regions of the universe in which the direction of causation is reversed relative to how it is around here. Then what direction does time have in those regions? Again, it seems that the reductionist must either say that time has no direction if there are regions such as this, or that time has different local directions.

There are also other problems with reducing the direction of time to the direction of causation. The most pressing of these is that we often distinguish causes from effects by appealing to their temporal order. But we cannot do that if we wish to use the causal order as the base asymmetry for temporal direction. So we need a way to determine which of two causally related events is the cause, and which the effect, without appealing to the temporal order of the events. This has proven non-trivial, and represents another difficulty for this kind of reductionism about direction.

5.6.3. Deliberation

The final view we will consider is one that aims to reduce the direction of time to the asymmetry of one or other psychological phenomenon. The particular version of the view we consider here is one according to which the direction of time is reducible to certain features of our deliberative systems. We have already noted that beings like us deliberate about events towards what we call the future, but do not deliberate about events towards what we call the past. One might try to explain this deliberative temporal asymmetry by supposing that time itself has a direction. The direction of deliberation aligns with time's direction. The proposal under consideration here, however, attempts to invert the order of explanation. Rather than explaining why we deliberate towards the future and away from the past in terms of time's direction (or in terms of some further temporal asymmetry such as causation or knowledge), instead the aim is to reduce time's direction to the direction of deliberation. How does such an account proceed?

The idea is that in order to be able to deliberate at all, each of us needs to divide the world up into the things that we take ourselves to be able to choose between (our options) and the things we take as fixed and immutable, and which we use as the basis for our deliberation. So, for instance, Sara takes it as fixed that tomorrow her kitchen will exist, and she will not be able to levitate, and that if she wants toast, she will need to use a toaster (since she won't be able to toast bread by looking at it). But Sara takes it as open that she can choose toast or cornflakes for breakfast (these are both options). Each of us can only deliberate about things we take to be open. Sara cannot, for instance, deliberate about whether to levitate tomorrow, given that she knows I cannot levitate.

The reductionist will say that the direction of time reduces to the direction of deliberation. That is, the future *just is* the direction *towards* which we deliberate, and the past *just is* the direction *away* from which we deliberate.

Again, though, two problems arise for this strategy. The first, and most obvious, is that in a mirror world we would expect agents in one temporal half of the world to deliberate towards what we call the future, and agents in the other temporal half to deliberate about what we call the past. So we should expect that in such a world a reductionist account of temporal direction will yield at least two different local temporal directions. A second problem is that it's not clear why we should think that the future is the direction towards which we deliberate, rather than the past being the direction towards which we deliberate. Even if deliberation is temporally asymmetric, and even if this shows that time is anisotropic, why does it show that time goes from what we call the past to what we call the future, rather than the other way around?

5.6.4. Returning to Reduction

What consideration of these reductionist stories suggests is that reducing temporal direction to some particular temporally asymmetric phenomenon will prove difficult. That's because the candidate base asymmetries are not always, everywhere, aligned in the same direction (entropy can decrease towards what we call the future; effects can precede their causes and so on), and because even if the candidate asymmetry is always aligned in the same direction, it's not clear that the mere presence of that asymmetry gives us reason to think that time is directed from what we call past to what we call future, rather than the other way around.

5.7. Summary

The asymmetry of time is one of the most perplexing physical features of our universe. Attempting to come to terms with asymmetries in time brings philosophy and science into close conversation. Fully grasping the kinds of asymmetries that there are, and understanding the options for explaining those asymmetries, is really a joint project in philosophy and physics. Our goal in this chapter has been to provide an overview of some of the central issues concerning the asymmetry, anisotropy and directionality of time. The key points covered were:

(1) The asymmetry, anisotropy and directionality of time must all be carefully differentiated from one another.

(2) There are a number of temporally asymmetric phenomena, including:

the asymmetry of causation, the asymmetry of entropy, the asymmetry of deliberation, the knowledge asymmetry and the directionality of time.
(3) The laws of physics appear to be temporally symmetrical, and so it is difficult to see how one might attempt to explain the various temporally asymmetric phenomena that there are in terms of lawful asymmetry.
(4) Dynamic theorists of time reduce temporal direction to the direction of temporal flow, and then attempt to explain the asymmetries of time by appealing to time's having a direction.
(5) Reductionists about temporal direction attempt to reduce direction to temporally asymmetric phenomena, or to temporal anisotropies (intrinsic or extrinsic).
(6) Statistical mechanics plus the assumption that the universe began in a state of low entropy may provide an explanation for the temporal asymmetry of entropy.
(7) Causation and the asymmetry of deliberation present alternative options for reducing temporal direction.

5.8. Exercises

i. Think of all of the physical asymmetries you can (whether to do with time or not). Make a list. Are any of these related to the asymmetries discussed in this chapter? If so, how?
ii. Describe a very basic physical system, such as the motion of a pendulum. Draw a sequence of diagrams that represent different temporal stages of this physical system. What changes, if any, do you need to make to these diagrams in order to represent the temporally reversed version of this physical system?
iii. Develop your own reductionist account of the direction of time. Compare it to one of the reductionist accounts that we have discussed here. Is your account better or worse than these accounts?
iv. Are all of the dynamic theories of time equally able to explain asymmetries in time by appealing to temporal flow, or is one of the dynamic theories better suited to providing such an explanation? Try to justify your answer.
v. Break into groups and invent a short play. Have one group perform the play forwards, and have the other perform the play backwards. Document the kinds of changes that you need to make in order to reverse the play in this manner.

5.9. Glossary of Terms

Base Asymmetry
The asymmetry to which the direction of time might be reduced.

Entropy
The measure of disorder of a system.

Extrinsic
A feature that something has in virtue of bearing certain relations to other things.

Grounding
The dependence of one thing on another.

Intrinsic
A feature that something has in virtue of the way it, itself, is, and nothing else.

Macrostate
The state of the properties of a system, such as temperature.

Microstate
The state of the particles that make up a system.

Primitive
Not explainable, or reducible to anything else.

Reduction
The identification of one thing with another.

Reductive Base
The thing Y that some X is reduced to (i.e. identified with).

Temporal Anisotropy
The asymmetry of time itself.

Temporal Asymmetry
The asymmetry of phenomena in time.

Temporal Directionality
The direction that time has, if indeed it has one.

Time-Reversal Invariance
The laws of nature work equally well in one temporal direction as they do in the other.

5.10. Further Readings

O. Shenker and M. Hemmo (2011) 'Introduction to the Philosophy of Statistical Mechanics: Can Probability Explain the Arrow of Time in the Second Law of Thermodynamics?', *Philosophy Compass* 6 (9): 640–51. Though not introductory, this is a relatively accessible overview of the connection between statistical mechanics (and hence entropy) and the arrow of time.

J. J. C. Smart (1953) 'The Temporal Asymmetry of the World', *Analysis* 14 (4): 79–83. This is a relatively early statement of the puzzle of temporal asymmetry, and of the distinction between a direction to time itself and a direction in the content of things in time.

H. Price (1996) *Time's Arrow & Archimedes' Point: New Directions for the Physics of Time* (Oxford University Press). A very thorough and generally accessible introduction to the physics, and in particular the apparent direction, of time.

J. Ismael (2016) 'How Do Causes Depend On Us? The Many Faces of Perspectivalism', *Synthese* 193 (1): 245–67. This is not an introductory work, but it is an accessible overview of perspectivalism about causation, which itself provides a nice entry into thinking about perspectivalism about temporal direction.

6

Time and Causation

So far we have considered a wide range of issues to do with time. In the background of much of this discussion has been the concept of causation. Indeed, for each of the main areas of philosophical interest in the notion of time, causation has some role to play. In this chapter we turn our attention to understanding this intriguing notion. We will draw a rather broad (and somewhat artificial) distinction between two very general theories of causation, before looking at the relationship between causation and time.

6.1. Two Theories of Causation

The two major theories of causation discussed in contemporary philosophy are process theories of causation and counterfactual theories of causation. We will have more to say about both theories in due course; for now, however, it is useful to place these two theories on a kind of 'metaphysical spectrum', from the most substantive to the least substantive theories of causation.

At one end of the spectrum we have process theories. These are the most substantive theories of causation in so far as they treat causation as a kind of physical process that is available for empirical investigation in the ordinary way. Process theories are substantive because they treat causation as something to be discovered, just like any other thing that we might seek to investigate scientifically.

At the other end of the spectrum we have counterfactual theories of causation. These theories are the least substantive in so far as they don't treat causation as any kind of physical process. Indeed, it is compatible with a counterfactual theory of causation that causation obtains in the complete absence of any physical process whatsoever. On this view, causation is analysed chiefly in terms of the truth or falsity of counterfactual conditionals: claims of the form 'if X had not been the case, Y would not have been the case'.

This way of carving up the space of theories of causation has one chief limitation: it doesn't cover all of the theoretical approaches to causation that there are. The advantage of dividing the terrain into process and counterfactual theories is that it produces a fairly straightforward narrative that helps us to see a lot of what is at stake in debates over causation. Because the truth should never get in the way of a good story, we will keep the division, but with the understanding that it is not supposed to be a completely accurate representation of the philosophy of causation.

6.2. Process Theories of Causation

Process theories of causation have a fairly long tradition in the history of Western thought. Contemporary process theories of causation have their roots in mechanistic theories of causation, which rose to prominence in the wake of the unprecedented scientific success of Newton's *Principia Mathematica*. For the first time in history, it seemed that humanity had tamed the heavens. The motion of the stars and the planets in our solar system had been given a detailed and precise mathematical characterisation that produced predictions of incredible accuracy. The picture of the universe that Newton paints is like the inner workings of a clock, with every gear and lever set on the tension of every other, to produce an exact and unrelenting process. An account of what it is for one thing to cause another follows very naturally from this picture of the universe.

6.2.1. Early Mechanism

One of the chief insights of Newton's picture is that what we really need to explain is deformations in motion. Everything in the universe is continuing on some path or other, and will do so unless impinged on by an external force. The concept of causation – which appears to be the very engine of this kind of change – is mechanistic in nature. According to this very basic mechanistic account, causation is – to put it somewhat crudely – a matter of things bumping into one another. Causation occurs when an object that is minding its own business, heading through the universe in a straight line, is struck by something and is then forced to change its motion in some respect.

In other words, according to a mechanistic theory of causation, P causes Q when P brings about a change in Q's motion. Causation, then, just is the alteration of the motion of one thing by another. Mechanistic theories of causation such as this have clear limitations. Chief among these is the fact that a great deal of causation has nothing to do with motion. When Sally touches a hot stove and burns herself, the burn that Sally receives is not a result of motion. It is, rather, the result of the transference of energy. Similarly, when the path of an asteroid is altered in virtue of the gravitational field of a nearby sun, it is not because the motion of the sun induced a change in the motion of the asteroid. It is, rather, the unmoving gravitational field of the sun that changes the asteroid's direction.

Mechanistic theories of causation, then, are no good. But they do preserve a core insight, one that many process theories seek to develop. That insight is that causation is an empirical phenomenon. Motion can be measured and detected, as can changes in motion. Inertia is well defined and understood, as is velocity. All of these are straightforwardly encoded in a working mechanics, such as Newton's. Causation, then, is within the purview of empirical investigation.

6.2.2. Recent Process Theories

Recent process theories of causation take this core insight and develop it further. Just as early process theories had their roots in the best mechanics of the time – Newton's – contemporary process theories have their origins in the best mechanics of our time – Einstein's. The basic idea behind contemporary process theories is to treat causation as a matter of objects in constant motion interacting in some manner. Because, in Einstein's theory, every object can be considered to be in motion relative to some frame of reference, every object has the capacity to exert a causal influence.

According to one prominent process theory, a causal process is a causal interaction between two persisting objects, and a causal interaction involves the transference of a 'mark' from one object that is persisting in this manner to another. We can think of a mark transfer as a local modification of one object, by another. Suppose that a chalk-covered white billiard ball hits a red billiard ball, leaving a white chalk mark on it. That is a paradigmatic instance of mark transfer: we can see the mark – the chalk – that is transferred from one object to the other.

Contemporary process theorists take the idea of mark transfer and provide a concrete account of what marks are, and thus what mark transference involves. According to contemporary process theories, a mark is a conserved quantity. Mark transference, then, is the transference of some conserved quantity or other. A conserved quantity is any quantity that obeys the laws of conservation. The contemporary process theorist relies on our best science to tell us what conserved quantities are. Candidates, however, include mass, energy, momentum and electric charge.

The contemporary process theorist provides a two-part definition of causation. First, they define a causal interaction as follows:

CQ1 A *causal interaction* is an intersection of world lines which involves exchange of a conserved quantity.

Then they define a causal process in terms of conserved quantities:

CQ2 A *causal process* is a world line of an object which possesses a conserved quantity.

Note that a 'world line' is just a trajectory through spacetime: a path from one spacetime point to another, where any two adjacent spacetime points in the path are light-like or time-like separated from one another (they are at a null or negative spatiotemporal distance to each other and thus one can get from one spacetime point to the other by travelling at a speed that is at or below the speed of light).

The basic idea is that causation occurs when two objects meet in spacetime (this is what happens when their world lines intersect), and one object transfers some conserved quantity to the other. Since there are a variety of conserved quantities, this idea can account for a wide variety of causal processes, from burning one's hand on a hot stove to one billiard ball causing another to change direction (or to be covered in chalk). And notice that the account does all of this in a scientifically upstanding and empirically tractable manner. Our best mechanics is used to define causation, and then causation is analysed in terms of empirically discoverable features of the world; features that obey important scientific laws (the conservation laws).

6.2.3. Causation by Omission

Unfortunately, as we shall now see, all process theories face a serious problem. Suppose that Sara buys a new plant. She affectionately names the plant Bertie. Sara tends to her plant dutifully, showering it with love, attention and, most importantly, water. One day, however, Sara goes on holiday and forgets to organise for someone to stop by and water her plants. Sadly, Bertie dies. It seems very plausible to say that Sara caused

Bertie's death by failing to water her beloved plant. And yet, every process theory of causation currently on the market has trouble delivering that straightforward result.

Why? Well, let's go through the reasoning. According to the contemporary process account outlined in the previous section, causal interactions occur only when the world lines of two objects meet and, when they do, a conserved quantity is transferred from one object to the other. So for Sara to cause Bertie's death at, say, spacetime point P – the exact time and place of Bertie's final breath – Sara would need to be located near enough to P to interact with Bertie. But suppose that Sara has gone on holiday to the USA and Bertie lives (and dies) in Australia. Well, then Sara and Bertie's world lines do not meet at P and so Sara can't cause Bertie to die. More generally, there just is no conserved quantity that is transferred from Sara to Bertie and so no causal process linking the two.

The problem, in its most general form, is this. Sara's failure to water Bertie is not, itself, an event, or a physical phenomenon. It is, rather, the *absence* of an event, or the absence of a physical phenomenon. Bertie's death is, however, a physical phenomenon: it is a spatiotemporally located event. According to all versions of the process theory of causation, in order for there to be causation between Sara and Bertie there must be a physical process of some kind linking an absence (Sara's failure to water) with a presence (Bertie's desiccated corpse). But processes don't connect absences with presences. They always connect existing physical phenomena, one to the other. So it would seem that there simply cannot be a physical process connecting the cause with the effect in this case. It follows that no process theory can do any better than the account outlined in the previous section. All such theories will fail to account for the causal relationship between Sara and her beloved Bertie.

The process theorist has a number of responses available to her to solve the problem. None of them seem very good. First, the process theorist might simply bite the bullet. Yep, Sara does not cause Bertie's death. So what? To see why this type of response won't do, it suffices to consider a structurally analogous case. Consider Sam. Sam has a child he affectionately names 'Rudolph'. Sam tends to his son dutifully, showering him with love, attention and, most importantly, food. One day, however, Sam goes on holiday and forgets to organise for someone to look after Rudolph and feed him. Sadly, Rudolph dies. It seems very plausible to say that Sam caused Rudolph's death by failing to feed his beloved son. And yet, every process theory of causation currently on the market has trouble delivering that straightforward result.

Now, suppose that the process theorist tells us to simply bite the bullet on this case. Yep, Sam is not a cause of Rudolph's death. So what? Suppose, further, that Sam is arrested and charged and brought before the judge for his negligence. If the process theorist is right, then Sam did not cause Rudolph's death. But on the plausible assumption that it is necessary for being *responsible* for an event *E*, that one is at least a *partial cause* of *E*, then it follows that Sam is not responsible for Rudolph's death – not morally, and not legally. That's preposterous: of course Sam is responsible for Rudolph's death (just try the 'but the process theory of causation is true' defence in a court of law and see how far it gets you!). But if Sam is responsible for Rudolph's death, then he must be causally implicated in that death somehow. The process theorist's denial that Sam is causally involved in the death looks troubling indeed.

One might take a 'divide and conquer' approach to this type of problem. The process theory of causation, the process theorist might argue, is a theory of causation that is fit for service in scientific contexts only. It does not work outside of very particular empirical situations, involving the transference of conserved quantities. To be sure, there is some other notion of causation that is broader than this purely scientific one, but the process theory is just not in the business of analysing that broader concept. The approach is a 'divide and conquer' approach because, in essence, the process theorist is recommending the sub-division of the concept of causation into two distinct concepts: one fit for service inside a legal context, and one fit for service inside a scientific context.

Whether this divide and conquer response works depends a bit on what it is that we are trying to do. If we are trying to give an account of causation regardless of the particular context in which a causal notion is used, then the process theory is not up to that task. If, however, we are trying to give an analysis of a peculiarly scientific concept of causation, then perhaps the process theory will do. Our view is that the goal of developing a theory of causation is, or at least should be, the more general one of analysing causation wherever it arises. The more localised project is not really a project of understanding causation. It is, rather, a project of understanding something like physical interaction inside science, or empirical influence as constrained by conservation laws or something along those lines. That project is of limited philosophical interest. The more interesting project, and the one that will really help us to understand what causation is, is the more general project. So it is a rather severe limitation of the process theory if it is just not up to the task.

The next option available to the process theorist is to reify absences. What does this mean? Well, recall the general form of the problem facing the process theory: the failure to water Bertie is the absence of an event; Bertie's death is not. Processes can't link presences to absences. But presumably that's because absences are not a part of the inventory of the universe. There is no 'thing' that corresponds to the failure of Sara to water Bertie. But what if there were such a thing? What if the failure was as much a spatiotemporally located object as was the death? In that case there would be no problem in linking the two things via some physical process or other.

This response to the general problem is costly. For it requires adding peculiar entities – absences – into our inventory of what exists. Such things cry out for a metaphysical account of their nature. But providing such an account looks hard indeed. Moreover, even if such an account could be given, the resulting theory would offend against parsimony in the extreme. For there is an infinite number of things that *fail to exist*, or to obtain. Once we open the door to some of these, it is difficult to see why we shouldn't countenance them all. Our account of the world would become bloated with these new objects. Worse still: once we have let these absences into the world, how do we stop them from being causally efficacious? For consider, it is not just Sara's failure to water the plants that brings about Bertie's death. Everyone on Earth failed to water the plants as well. If all of these various failures are *things* that *exist* it is difficult to see why they shouldn't also be causally implicated in Bertie's death. But that seems bizarre. Consider again the case of Sam and Rudolph. Everyone on Earth failed to feed Rudolph. So should we all go to prison? Even the judge sitting on the case? No; something has gone wrong.

The process theorist is not done yet. She might deny that absences are things, but nonetheless maintain that there is some process implicated in the relationship between Sara and Bertie. Think of it this way: there is a range of physical processes that bring it about that Bertie dies. We should think of the absence of watering not as the lack of a process, but as the initiation of a process which eventually kills Bertie. The process, in this case, being the process of desiccation. So while it seems natural to say that it is Sara's failure to water Bertie that kills him, what actually occurs is that Sara sets another process off – a process of heating by leaving Bertie in a hot room, or a process of wilting that is brought about by leaving Bertie in the sun – and so on, and it is these processes that link Sara to Bertie's death. In short, talking of Sara's failure to water Bertie as the cause of his death is a metaphorical way of identifying some underlying process that Sara does, in fact, bring about.

The trouble with this solution is that the general problem under consideration can be reconstructed for these other processes. Without going into details, just about every physical process involves two things at some point: excitation, whereby something is forced to happen, and inhibition, whereby something is prevented from happening, and this prevention in turn brings something else about. So when we heat the room in which Bertie lives, this heating in turn inhibits certain things from happening inside Bertie, such as the process of cellular respiration inside Bertie's cells. The lack of cellular respiration then brings about Bertie's death. In order to fully specify the process that is responsible for Bertie's death, a process that Sara initiates, we will be forced to once again call upon causation by some absence or other inside the relevant process that we have appealed to.

Now, we could try and make the same move with this second case of causation by absence, namely: find a third process to reinterpret the causation by absence, and treat the causation by absence as a kind of metaphor. But right down to the sub-atomic level we find processes of inhibition occurring. So it is very doubtful that we can find a scale at which, for every case of causation by an absence, there is a process that fully accounts for the causation in the relevant situation and that features no inhibition of any kind. In short, the universe seems to feature causation by absences at the most fundamental levels, making it very difficult to treat all statements of causation as metaphorical ways of specifying some underlying process.

This exhausts the standard solutions to the problem of causation by absence facing the process theory. In what follows we turn our attention to counterfactual theories of causation which, among their many virtues and vices, manage to provide an easy solution to the problem of causation by omission.

6.3. Counterfactual Theories of Causation

As we have already noted, modal claims deal with possibility and necessity. A counterfactual is a unique kind of modal claim. The easiest way to gain a sense of what a counterfactual is, is to look at some examples. So consider the following:

[1] If Trump had not won the 2016 US election, Clinton would have.
[2] If the 2016 referendum in the UK had received a majority of remain votes, the UK would not have decided to leave the European Union.

A counterfactual tells us what would or might have happened, had something that in fact occurred, not occurred; or if something that did not occur, had occurred.

A counterfactual theory of causation seeks to build causation around claims such as the above. Such theories attempt to analyse the causal relationship between distinct events in terms of counterfactual relationships between claims about those events. We start by defining a notion of counterfactual dependence as follows:

> **Counterfactual Dependence:** A counterfactual depends on B just when [i] if it were not the case that A, then it would not be the case that B, and [ii] if it were the case that A then it would be the case that B.

To understand this definition it is useful to consider a very basic example. Suppose that Sara strikes a match and lights it. Now consider the two events: E_1, Sara striking the match, and E_2, the match lighting. E_2 counterfactually depends on E_1 when two counterfactual claims are true, namely:

[3] If E_1 had not occurred, then E_2 would not have occurred.
[4] If E_1 had occurred, then E_2 would have occurred.

We know that the second counterfactual is true, very roughly, because we can look to see what actually happened. Given that Sara struck the match and her striking the match was followed by a lighting, we have good reason to suppose that [4] is true (we will sharpen this up in due course). The first counterfactual is more difficult to evaluate. But here's the basic idea. First, we imagine a situation that is just like the world we live in, except that in that world Sara doesn't strike the match. Based on what we know about science and the laws of nature, we then run the scenario forwards in our minds to see what happens with the lighting of the match. If the match does not light, then we have reason to suppose that [3] is true. If the match lights anyway, then we should think that [4] is false (this will also be made more precise in a moment).

We can use the basic notion of counterfactual dependence outlined above and formulate it into a first-pass theory of causation, as follows:

> **CF Theory of Causation:** 'X causes Y' is true iff there is a causal chain leading from X to Y consisting of events *A*, *B*, *C* ... such that *C* depends counterfactually on *B*, *B* depends counterfactually on *A* and so on.

Here's the basic idea. Consider two events, E_1 and E_2. E_1 causes E_2 when there is a chain of events linking the two such that any two adjacent links in that chain stand in a relation of counterfactual dependence. So consider, again, Sara and her match lighting. As before, let E_1 be the event of Sara striking the match and let E_2 be the event of the match's lighting. In between E_1 and E_2 there is a sequence of events:

A: The match head heats up.
B: The material on the match head combusts.
C: The match head catches fire.

E_2 counterfactually depends on C: if the match head had not caught fire, then the match would not have lit. Similarly, C counterfactually depends on B: if the material on the match head had not combusted, the match head would not have caught fire. B, in turn, counterfactually depends on A: if the match head had not heated up, then the material on the match head would not have combusted. Finally, A counterfactually depends on E_1: if Sara had not struck the match, the match head would not have heated up.

Using this basic model, it is easy to extend the counterfactual theory of causation to handle a great many of the cases of causation in which we are interested. One of the important features of the counterfactual theory is that it can handle all of the cases of causation invoked by the process theory of causation, and then some. Processes of the kind that process theorists are interested in can be modelled using chains of counterfactual dependence, where each step in the chain specifies a particular event – a localised happening in spacetime – and where any two spatiotemporally adjacent steps stand in a relation of counterfactual dependence.

But while the counterfactual theory of causation can handle instances of process-based causation, it does not demand the existence of processes linking cause and effect for causation to occur. And therein lies the chief advantage of the counterfactual theory: it is flexible enough to handle causation no matter what in the world underlies the causal facts in question. Because of this the counterfactual theory of causation is capable of handling causation by absence with ease.

To see this, suppose, once again, that Sara goes on holiday and neglects to water her beloved Bertie. To fit this case into a counterfactual theory of causation, we begin by identifying the sequence of events that leads from Sara's failure to water Bertie to Bertie's unfortunate demise. We can imagine that the sequence of events goes something like this:

E$_1$: Sara fails to water Bertie.
A: Bertie becomes dehydrated.
B: Photosynthesis in Bertie's cells ceases.
C: Bertie's cells cease respiration.
E$_2$: Bertie dies.

Having identified the relevant sequence of events, we then chain the events together with relations of counterfactual dependence, like so:

[1] If Sara had watered Bertie, Bertie would not have become dehydrated.
[2] If Bertie had not become dehydrated, photosynthesis in Bertie's cells would not have ceased.
[3] If photosynthesis in Bertie's cells had not ceased, Bertie's cells would not have ceased to respire.
[4] If Bertie's cells had not ceased to respire, Bertie would not have died.

Because these four counterfactuals are true, there is a causal chain leading from E$_1$ to E$_2$ of the kind required by the counterfactual theory of causation. Because there is a chain of the relevant kind, it follows that Sara caused Bertie's death.

One of the nice things about the counterfactual theory of causation is that it treats causation by absence in just the same way as it treats any other kind of causation. In every case a chain of counterfactuals is used to establish a causal chain linking cause with effect, a causal chain that underwrites causation.

It would seem then that the counterfactual theory of causation has a lot to recommend it. Before looking at some of the difficulties that the counterfactual theory of causation faces, it is important to dig down into the details of the view a bit more. In particular, it is important to say a bit about how we determine whether a counterfactual conditional is true.

6.3.1. The Semantics of Counterfactuals

As we have seen, a counterfactual is a kind of if/then statement: it is a conditional. But it is not like the usual kind of if/then statement that we encounter in a first-year logic course. That kind of if/then statement has the following truth conditions: the statement 'if A then B' is true just when either A is false, or B is true.

The truth conditions for a counterfactual conditional are more complicated. To understand the conditions, we need the concept of a possible world. Recall that a possible world is just a way that the entire universe could be: it is a complete specification of a possibility. For present purposes it will be useful to imagine our universe floating in a sea of bubbles, where each bubble is a self-contained cosmos that differs from our universe to greater or lesser degrees. Each such cosmos is a possible world.

The next thing we need is the concept of a similarity ordering. In particular, we need the idea that possible worlds can be ordered with respect to the actual universe in terms of how similar those worlds are to the actual universe. The most similar worlds to the actual world are the closest possible worlds. The most dissimilar possible worlds to the actual world are the furthest possible worlds.

Using these two notions, we are now in a position to state a first-pass semantics for counterfactual conditionals. Here it is:

Analysis 1: For any counterfactual of the form 'if it were that A, then it would be that B', that counterfactual is true just when some possible world in which both A and B are true is closer to the actual world than any possible world in which A is true and B is false.

To see how Analysis 1 works it is useful to apply it to a particular case. Suppose, again, that Sara strikes a match and the match lights. Now, consider the counterfactual dependence of the lighting of the match on Sara's striking, as enshrined in the following conditional:

[5] If Sara had not struck the match, the match would not have lit.

[5] is true when there is some possible world in which Sara does not strike the match and in which the match does not light that is more similar to the actual world than any world in which Sara does not strike the match and the match lights anyway. The thought, then, is that the counterfactual is true when it is more of a departure from actuality when Sara does not strike and the match lights anyway than when Sara does not strike and the match does not light. Intuitively, the counterfactual is true: if we imagine a scenario in which Sara does not strike the match, then, assuming that the world is exactly like our world in all other respects leading up to the striking, there won't be anything else available to light the match in lieu of Sara's striking. So the match won't light.

6.3.2. Pre-emption

So far we have looked at a basic counterfactual theory of causation, and have provided an overview of the semantics for counterfactual conditionals. Let us now consider the chief problem facing a counterfactual theory of causation. The central problem for a counterfactual theory of causation concerns a phenomenon called *pre-emption*. Pre-emption tends to occur when, for some causal chain, there is a back-up system in place, ready to kick in if the main causal chain fails. There are many ways to formulate a problem based on pre-emption, but we will focus on just one to give a flavour for the difficulties that back-up systems pose.

Suppose Sara and Emma are throwing rocks at a bottle. Sara throws, and then Emma throws, just after Sara does. Sara's rock gets there first at the last second, striking the bottle and smashing it. Emma's rock whizzes through the debris field. But for the accuracy of Sara's throw, Emma's would have done the job.

It seems clear that Sara's throw caused the smashing of the bottle. Unfortunately, the counterfactual theory of causation cannot deliver that result. In order for it to be the case that Sara caused the smashing of the bottle, there must be a chain of events leading from Sara's throw to the smashing such that any two adjacent events in that chain stand in a relation of counterfactual dependence. Let us model the chain, roughly, as follows:

E_1: Sara throws the rock.
A: Sara's rock flies through the air towards the bottle.
B: Sara's rock strikes the bottle.
E_2: The bottle smashes.

For Sara to be the cause of the smashing, each of the counterfactuals in the following sequence must be true: if Sara had not thrown the rock, Sara's rock would not have flown through the air towards the bottle; if Sara's rock had not flown through the air towards the bottle, Sara's rock would not have struck the bottle; if Sara's rock had not struck the bottle, the bottle would not have smashed.

The last counterfactual in this sequence is false, because of Emma's throw. It is just not true that if Sara's rock had not struck the bottle, then the bottle would not have smashed. That's because if Sara's rock had somehow missed the bottle, Emma's rock would have still been on target and would have hit the bottle. In other words, it is not true that some world in which Sara's rock does not strike the bottle and the bottle does not smash is closer than any world in which Sara's rock does not strike the bottle and the bottle smashes anyway. That's because, in the relevant world in which Sara's rock does not strike the bottle, Emma's rock strikes the bottle instead, and the bottle smashes regardless.

So the counterfactual theory yields the result that Sara does not cause the bottle to break. The problem is quite a general one: whenever there is a back-up system of the kind just described, the chain of counter-factual dependencies needed to establish causation is undermined by the back-up.

Now, you might think that the problem posed by pre-emption is not all that deep. After all, there is a physical process linking Sara's throw to the smashing of the bottle, and there is no such process linking Emma's throw to the smashing of the bottle. It might be thought, then, that there is scope for a proponent of the counterfactual theory of causation to avoid the problem by leaning on some of the resources made available by the process theory of causation. But there are two problems with this response. First, when the counterfactual theorist makes use of the resources available to the process theorist she opens herself up to the problems posed by absence causation (just imagine the Sara and Emma story recast in terms of causation by absence). Second, even if the counterfactual theorist can somehow make use of the process theorist's picture of causation to solve the problem posed by pre-emption, there are nastier versions of the problem that can be formulated. Can you think of one?

6.4. Causation and Time

We have before us two prominent theories of causation: process theories of causation and counterfactual theories of causation. In this final section we will turn our attention to the relationship between causation and time.

6.4.1. Retro Causation

Let us call causation that goes backwards in time 'retro causation'. As we shall see later on, time travel scenarios seem to require backwards in time causal influences. For instance, suppose that Tim steps into a time machine on Tuesday and steps out of the time machine on the previous Monday. Tim's stepping into the time machine causes his exit from the time machine. His entrance into the time machine, however, occurs after his exit. So the causal influence in virtue of which Tim brings it about that he travels through time must be moving in the future-to-past direction, which is the opposite direction from the one in which causation typically moves.

That time travel requires retro causation is undeniable. What is less clear is how the two theories of causation considered thus far fare with respect to accommodating causation of this kind. At first glance, it seems that the counterfactual theory of causation has an advantage over the process theory of causation. As we have already seen, the counterfactual theory of causation requires no physical processes to underlie causal influence. Its ability to easily handle causation by absence is a testament to this fact. This seems to be a benefit in the present context because backwards causation, if cast in process terms, would seem to require a process that goes backwards in time.

The trouble is that such backwards in time processes appear to be unlikely. Even in a fully relativistic spacetime, the background metric structure of the universe seems to have a causal direction built into it. This is not to say that backwards in time processes are impossible. They are clearly possible. There are solutions to Einstein's field equations (the equations that underlie general relativity) that permit processes of this kind. But even though such processes are possible, the universe has to have a very particular structure in order for such processes to arise.

That backwards in time processes are unlikely is not in and of itself a problem. The problem is that the process theory of causation defines causation in process terms. So the process theory of causation has built into it the idea that backwards in time processes are unlikely. The counterfactual theory of causation, by contrast, does not seem to have the same implication, since it does not tie causation so directly to processes.

Ultimately, however, both the process theory and the counterfactual theory have the same scope – limited as it is – to accommodate backwards in time causation. Both the process theory and the counterfactual theory tie causation to the laws of nature. The process theory does this indirectly, by tying causation to processes that are, presumably, law-governed. The counterfactual theory does this directly, by using the laws of nature as a basis for evaluating counterfactuals. As discussed in Chapter 5, the laws of nature display deep symmetries when it comes to time. These deep symmetries suggest that, in so far as the laws of nature are concerned, causation should be possible in both the past-to-future direction and the future-to-past direction. Which is to say that the laws of nature provide no in principle reason to suppose that retro causation can't happen.

Moreover, in so far as there is a temporal asymmetry to be found, it seems to be more to do with the distribution of matter and energy within spacetime which, in turn, is a function of the past hypothesis: the idea that the boundary conditions of the universe display low entropy. It is plausibly because of this distribution, combined with statistical mechanics, that we see the types of asymmetries that we do. More carefully, in so far as processes that go backwards in time are unlikely, it is because of statistical mechanics in combination with the past hypothesis. But these rather global constraints that make backwards in time processes unlikely really extend to all causal influences; it really has nothing to do with physical processes per se. Causal influences that go against the entropic grain of the universe, as it were, are unlikely. This low probability affects the process theory and the counterfactual theory of causation equally. In both cases, the laws of nature plus the boundary conditions of the universe make retro causation a difficult affair. Ultimately, then, neither of the two theories that we have looked at boasts much of an advantage when it comes to accommodating the kind of causal influences needed for time travel.

6.4.2. Is Time Reducible to Causation?

Is time reducible to causation? There are really two distinct issues one might have in mind in asking this question. One is the broad question of whether we can reduce temporal relations to causal relations. The second is a narrower question of whether we can reduce temporal directionality to causal directionality. We have already considered this second question in Chapter 5, so we will not further consider that issue here. Instead, we turn to the broader question of whether temporal relations (i.e. time itself) might be reducible to causal relations (i.e. causation).

First: what it would it take to reduce time to causation? Well, presumably, we would need to take a particular theory of what causation is and, with respect to that theory, show that time itself can be identified with whatever causation turns out to be. Achieving a reduction of this kind is going to be easier if we use a counterfactual theory of causation. That's because the process theory of causation, at least as that theory has been outlined here, presupposes a spatiotemporal structure as the backdrop against which causation is defined. Which is to say that causation is analysed partly in terms of time. If we were to then try and reduce time down to causation, the resulting reduction would result in a deep and unpleasant circularity: causation exists because time exists, and time exists because causation exists. Not all circles are bad, but this one looks unpleasant. For it is an explanatory circle: time partly explains causation and causation explains time. This violates plausible constraints on explanation, namely that explanation is not symmetrical in this manner.

Even if such a reduction is possible, why would anyone want to reduce time to causation in this fashion? We can see at least three reasons. The first is an appeal to Ockham's Razor. If we can reduce time to causation, then we have fewer aspects of the universe to try and explain. In particular, if we can reduce time to causation then we might be able to use whatever explanations we have available of causal asymmetries – the fact that retro causation seems unlikely, while ordinary causation is commonplace – to explain temporal asymmetries, thus reducing the explanatory tasks we are required to complete by one.

Second, by reducing time to causation in this manner we may gain a better understanding of what time is and how it works, an understanding that can then be brought back into physics. For instance, if time is reduced to causation, then that would seem to fit neatly with the causal sets approach to quantum gravity discussed in Chapter 4. Even if we can't fit the resulting picture of time and causation into an existing theory, we may nonetheless still gain some insight into how time works, since it will inherit whatever properties causation has.

Third, the reduction of time to causation brings with it the potential to provide an explanation of temporal flow. As we saw when discussing the dynamic theory of time and the various difficulties it faces, there are many different ways of characterising the flow of time. What we didn't really consider, however, is that some of these ways of understanding the flow of time stand in need of explanation. Consider, for instance, the flow of time as it is understood within the growing block theory of time. According to the growing block theory, the flow of time involves the gradual coming into existence of new slices of reality. But exactly what is the mechanism behind this coming into existence of new slices? What makes the flow of time happen? The reduction of time to causation presents us with a potential answer to this question. The process just considered, whereby new slices of reality come into existence, is a causal process, and so as a process it is just like any other case in which some entity is brought into existence via a causal influence. Of course, whether or not this is a viable picture of time overall remains contentious.

Ultimately, whether it is possible to reduce time to causation depends upon whether or not causation itself is grounded in, or based on, temporal notions. We need to use a theory of causation that does not, itself, build time into the analysis of causation. Since the counterfactual theory of causation doesn't build time directly into the way in which causation is analysed, it is more amenable to the kind of reduction under consideration. But if the counterfactual theory of causation doesn't build time into its analysis of causation, then this opens up an intriguing possibility: instead of reducing time down to causation, can we simply eliminate time altogether without thereby also eliminating causation? Can there be causation in a timeless world? Let us briefly consider this possibility.

6.4.3. Causation in a Timeless World

The chief challenge in disentangling time and causation is to give an account of causation that does not somehow presuppose temporal notions. In principle, at least, the counterfactual theory of causation looks like a good bet in this respect. As we have already seen, the counterfactual theory of causation analyses causal relations in terms of chains of counterfactual dependence. A great deal of the work in developing this theory, then, is being done by the underlying counterfactuals. Counterfactuals, however, are not essentially temporal. Indeed, there appear to be many counterfactuals that don't involve temporality at all, or at least do not depend on time for their truth or falsity. For example, consider the following counterfactuals:

> If the molecular structure of the diamond had not been a covalent network lattice, the diamond would not have had the same hardness.
>
> If the inverse square law of gravitation had been an inverse cube law, galaxies would not have formed.
>
> If the neural structure of Sally's brain were different, she would not have an experience of pain.

Each of these counterfactuals appears to be true, or at least plausible (perhaps subject to being worked out in a bit more detail). And yet while these counterfactuals are about things that happen in time, or things that exist in time, the actual truth of the counterfactuals does not seem to be essentially linked to time.

To see this, let us imagine a timeless universe. As we discussed in Chapter 4, we will suppose a timeless universe to be one that lacks even a C-series (and hence also lacks a B-series and an A-series). In such a universe there may be a number of *three-dimensional spatial configurations*. Each of these corresponds to a complete specification of the location of every particle that there is, plus all of the inter-particle distances. One could think of these (very roughly) as being each of the physically possible ways of distributing all the particles in our universe, at a time. Importantly, however, these spatial configurations are not connected by any temporal relations whatsoever. There is, as it were, no proper or correct way to order these configurations into a temporal ordering. No spatial configuration is earlier than another; no spatial configuration is later than another; nothing is simultaneous with anything else, since all spatial configurations are purely spatial: they have no internal temporal relations either. If you were to think of the spatial configurations as being like pieces of a jigsaw puzzle, then this is equivalent to the claim that there is no right way to fit these pieces together to form a unified picture.

Now, let us take one of these spatial configurations. In that spatial configuration, let us suppose, there is a diamond with a covalent network lattice. The first counterfactual above seems to be true of the diamond even though there is no time in the universe under consideration. If we were to imagine a similar timeless world in which the diamond in the relevant spatial configuration lacks a covalent network lattice structure, then it seems plausible that the diamond would not be as hard in that situation. Similarly, take the full range of spatial configurations and suppose that gravity across the relevant spatial configurations is governed by an inverse square law, or at least appears to be: in so far as there is any gravitational relationship between objects in space, the gravitational fields in a given spatial configuration always obey an inverse square law. Now, suppose we consider a sequence of spatial configurations in which gravity is obeying an inverse cube law. Again it seems plausible that the structure of the spatial configurations would be wildly different. It doesn't seem out of order to suppose that these spatial configurations won't feature any galaxies. Finally, take a spatial configuration in which Sally exists, and the neural structure of her brain is thus and so. Now, consider a nearby timeless world in which the neural structure of Sally's brain is different. Then it would be plausible to suppose that she would be having a different experience.

Now, we recognise that this last example (and perhaps the gravitational example as well) may seem a bit strange. If there is no time in the imagined world, how can be Sally be experiencing anything. Isn't time a precondition for experience? Perhaps. But that question is somewhat orthogonal to the point we are trying to make. The point is just this: there can be counterfactual relationships between low level phenomena – such as neural states or molecular structures – and high level phenomena that appear, in principle at least, detachable from temporal notions.

All we have shown thus far is that counterfactuals can be true even in worlds without time. That is a far cry from showing that a theory of causation can be given that does not presuppose time. It does, however, at least provide a glimmer of hope towards completing that larger project. For if counterfactual dependence itself is not essentially temporal, then there may be a way to analyse causation in terms of counterfactual dependence whereby the resulting analysis does not make essential use of time.

The chief challenge to developing a counterfactual theory of causation along these lines is to show that all of the cases of causation that we know and love – the types of things that motivate the development of a theory of causation in the first place – can be recovered using a notion of counter-factual dependence that is not essentially temporal.

6.4.4. Causation, Time and Direction Revisited

In this chapter we have asked whether we might be able to reduce time to causation, and in Chapter 5 we asked whether we might be able to reduce the direction of time to the direction of causation. Both of these questions, and our discussion of causation at the beginning of this chapter, reveal something important. We have largely assumed that causation itself is directed. That is, we have assumed not only that, typically, causes temporally precede their effects (causation is temporally asymmetric), but also that causation goes *from* cause, *to* effect. We very naturally think of causation as directed, in the same way that we naturally think of time as directed: as going *from* earlier, *to* later. If causation is, indeed, directed in this way, then the temptation to try to reduce some aspect of temporality to causation is inevitable: for we can hope to get the direction of time from the direction of causation. (Alternatively, we could try to get the direction of causation from the direction of time, by noting that causes are, by their very nature, the things that happen earlier in time than do effects.)

It is, however, worth noting that just as there are some who reject the contention that time has a direction – namely C-theorists – there are also those who reject the contention that causation has a direction. Though these are distinct views they are quite naturally paired together. Just as C-theorists think that, really, there is simply a temporal ordering of events, but no temporal direction to those events, the analogous view about causation is that there are causal connections between events, but no causal direction. We can think of the latter as the view that causation is like the glue between events. The glue is real: it really does hold events together. But there is no sense in which the glue goes from one event (the cause) to another (the effect). Still, just as C-theorists think there is a mere *appearance* as of time having a direction, an appearance that is, in part, the result of our particular local environment and our particular psychologies, so too on this view of causation there is an appearance as of causation being directed. And, again, this will likely be explained by our particular psychological features and our epistemic goals. In both cases the idea is that the world itself is entirely symmetrical with respect to time and causation. It is just that our particular orientation in the world is such that we experience it as asymmetrical: as directed from earlier to later, from cause to effect.

It is important to notice that even if causation is not directed, this does nothing to defang the objections to timeless theories that we have already encountered. For it does not immediately help to show how to make sense of causation in the absence of time, since it still seems that we need a C-series ordering of events to make sense of causation, even if causation itself is undirected. But it is a C-series ordering of events that we are lacking in a timeless world, and so the resources with which to generate an account, even an account of undirected causation, are still meagre.

6.5. Summary

In this chapter, we have provided a potted overview of the metaphysics of causation. There is a great deal more to say about causation than has been outlined here. Theories of causation are legion, and we have really only focused on two prominent theories. These two theories, and the debate between them, do however structure much of the contemporary discussion surrounding the metaphysics of causation. The key points we have covered in this chapter may be summarised as follows:

(1) Contemporary process theories analyse causation in terms of the transfer of conserved quantities between interacting objects in spacetime.
(2) Process theories of causation have difficulty accommodating causation by omission.
(3) Counterfactual theories of causation analyse causation in terms of chains of counterfactual dependencies.
(4) Counterfactual theories can handle cases of causation by omission, since counterfactual dependence does not require the existence of any physical processes.

(5) Counterfactual theories of causation have difficulty accommodating pre-emption, which involves the existence of causal back-up systems that undermine counterfactual dependence.

(6) Process theories and counterfactual theories of causation perform equally well when it comes to accommodating retro causation.

(7) Counterfactual theories of causation appear to be better suited to reducing or eliminating temporal relations.

(8) Causation, like time, may be undirected; there may be a causal analogue of the temporal C-series.

6.6. Exercises

i. Outline one version of the process theory of causation. Identify three problems for that theory. Try to find a solution for each problem.

ii. Consider the problem of causation by omission. Do you think that this is a serious problem for the process theory or do you think that it can be overcome?

iii. Break into three groups: the prosecution, the defence and the evaluation. Put the counterfactual theory of causation on trial. Let the prosecution argue that the counterfactual theory is false, let the defence defend the counterfactual theory, and let the evaluation declare a winner. Take it in turns to argue until a verdict has been reached.

iv. Can you see a way to modify the counterfactual theory of causation so as to avoid the problem of pre-emption? Defend your proposal.

v. Can you think of any other ways to order worlds in terms of similarity? Are any of these better than using our 'gut feelings'?

vi. Consider the idea that causation is undirected. Can you foresee any problems for this view of causation?

6.7. Glossary of Terms

Absence
The lack of an event.

Conserved Quantity
A quantity in physics that obeys the laws of conservation, e.g. energy.

Counterfactual
A modal claim about what would or might have happened had something

that actually occurred not occurred, or had something that actually did not occur, occurred.

Counterfactual Dependence
An event E counterfactually depends on an event E* when if E had not occurred E* had not occurred and if E had occurred E* would have occurred.

Ockham's Razor
A principle of methodology; the simpler theories are more likely to be true.

Pre-emption
A case of causation in which there is a back-up system operating.

Retro Causation
Causation of an event at t by an event at some time t+ after t.

Spatial Configuration
A complete specification of all of the particles in the universe at a time, including all inter-particle distances.

World Line
A trajectory through spacetime.

6.8. Further Readings

C. Hitchcock (2007) 'Three Concepts of Causation', *Philosophy Compass* 2 (3): 508–16. Hitchcock's piece provides a useful introductory overview of the debate surrounding causation. The paper is very 'big picture' and so is useful for getting the lay of the land before diving into more advanced material on causation.

S. Bernstein (2015) 'The Metaphysics of Omissions', *Philosophy Compass* 10 (3): 208–18. While this is not an introductory work, it is a fairly accessible overview of the complicated literature on the metaphysis of omissions.

D. Lewis (1973) 'Causation', *Journal of Philosophy* 70 (17): 556–67. This is not an introductory work, but it is a nice early statement of the counter-factual theory of causation.

7

Persistence Through Time

There is a very banal sense in which everything that is not instantaneous travels through time. Since that is not the interesting kind of time travel – the kind that requires a time machine, which we shall discuss in the next chapter – we call that kind of time travel we get by existing at multiple times *persistence*. In this chapter we examine a number of different accounts of persistence. These are rival accounts of what it takes for something to persist through time. We will evaluate these different accounts before moving on to consider, for each, whether it is consistent with all of the different views about temporal ontology considered in Chapter 1.

7.1. Endurance and Perdurance: an Overview

There are two main rival candidate theories of persistence: *endurantism* and *perdurantism*, or, as they are also sometimes known, *three-dimensionalism* and *four-dimensionalism*. As we will see in section 7.3, a number of other theories of persistence have also been developed. But before we consider those, we want to get a broad understanding of endurantism and perdurantism. Exactly how to spell out each of those views, and the difference between them, is a vexed issue, and one to which we will return. For now, our aim is just to give a rough characterisation of each so that we can see what is at stake in the debate.

Endurantism is the view that objects persist by enduring: by being *wholly present* at more than one time. Perdurantism is the view that objects persist by perduring: by being *partially present* at more than one time. Let's consider perdurantism first, since it will prove easier to get a handle on endurance if we can contrast it with perdurance.

To understand perdurantism let's begin by considering Annie the dog, at time t. Annie has extension through space at t. She takes up a region of space. It is plausible that she does so by having different spatial parts at different spatial locations. Why does Annie take up just that region?

Because she has feet in some part of that region (in some sub-region), and a head in some other part of that region, and a tail in some other part of that region, and so on and so forth. The intuitive thought is that Annie at t is just the sum of all of her parts at t, and that Annie is extended through space at t, by having different parts at different spatial locations at t. Here, it will be helpful for us to distinguish between parts, and proper parts. In mereology, 'proper part' is what is normally meant by 'part'. A proper part is some part of an object that is less than the whole object. So Annie's foot is a proper part of her, as is all of her minus her nose, and so on. All of Annie, by contrast, is a part of Annie, but not a proper part of Annie.

Perdurantists think that the way Annie is extended across space is the same as the way persisting objects (i.e. perduring objects) are extended across time. Perduring objects exist at different times by having proper parts present at each of those times. Perdurantists have introduced a special name for those proper parts: *temporal parts*. Temporal parts are short-lived objects. Indeed, the shortest lived such object is an instantaneous temporal part. This is the temporal part of a perduring object that exists for a single instant of time. That object is three-dimensional: it has extension only across the three spatial dimensions, and not along the temporal dimension.

According to the perdurantist, if you see an object, like Annie, for just an instant (t), then what you are seeing is an instantaneous temporal part of Annie: the thing made up of all of Annie's spatial parts at t. But that thing is not Annie, it is just a little part of Annie. Annie is the sum of *all* of those instantaneous temporal parts. So Annie herself is four-dimensional, because she is made up of a series of three-dimensional temporal parts. She is four-dimensional because she is, quite literally, extended along the temporal dimension the way that, at a time, she is extended along the three spatial dimensions. Looked at from a 'god's eye' perspective Annie looks like an Annie-shaped spacetime worm: she's long and thin (like a worm) along the temporal dimension, and at each time along that dimension she is Annie-shaped. This is why perdurantism is sometimes known as four-dimensionalism: because it is the view that ordinary persisting objects are four-dimensional, not three-dimensional, as we might have supposed.

What does all of this tell us about the relationship between Annie at one time, and Annie at another time? Well, in one sense it tells us that persisting objects are not numerically identical over time. Numerical identity is the relation that everything bears to itself and nothing else. So if X and Y are numerically identical, there is just one thing, and two names for that thing: X and Y (just as when we discover that Clark Kent is Superman, we discover that there is just one chap, and two names for him). In what sense are perduring objects not numerically identical across time? Well, to be sure perduring objects are self-identical. The four-dimensional worm that is Annie is identical with itself and with nothing else. Nevertheless, Annie at any one time – that is, the temporal part of Annie that exists at that time – is not numerically identical with Annie at any other time, that is, with the temporal part of Annie that exists at that other time. For no two temporal parts are identical (they are, after all, two not one). So the three-dimensional thing that we meet at one time is not identical with the three-dimensional thing that we meet at some later time, and in this sense, Annie is not numerically identical across time.

By contrast with perdurantists, endurantists hold that persisting things endure through time. Enduring objects, unlike perduring objects, are not four-dimensional: they are three-dimensional. That is why endurantism is sometimes known as three-dimensionalism. (Though notice that this is only an apt name in a world with three spatial dimensions: in a world with two spatial dimensions enduring things will be two-dimensional, and in a world with four spatial dimensions enduring things will be four-dimensional.)

The perdurantist, recall, thinks that there exists a series of very short-lived (indeed, instantaneous) three-dimensional objects – instantaneous temporal parts – and that persisting things are composed of all these objects. By contrast, endurantists think that each of these instantaneous three-dimensional objects just is the whole of the persisting thing. When you see Annie at t, you see *all* of Annie. Annie does not have temporal parts. So endurantism is the view that persisting objects are *multiply located in time*. That's because it's the view that all of Annie exists at t_1, and all of Annie exists at t_2, and all of Annie exists at t_3, and so on. To get a handle on this, think about properties for a moment. On some views of properties, the property of redness, say, can be *wholly instantiated* in Sara's coat, and in your socks. It is not that some *part* of redness is in Sara's coat, and some other part of it is in your socks: it is the same redness that is in both Sara's coat and your socks. The very reason that both Sara's coat and your socks look alike (in terms of colour) is that the property of redness is multiply locatable: it is located in both places. Endurantism is the view that persisting objects are like redness in this respect.

What is attractive about this view, according to endurantists, is that we are not committed to there existing a whole series of distinct instantaneous objects; instead, there is a single object, Annie, that exists at multiple times. Let's return to think about numerical identity for a moment. What is the relationship between Annie at t_1 and Annie at t_2? Well, they're both just Annie. Annie at t_1 is numerically identical with Annie at t_2. Suppose at t_1 Annie breaks into the trash can and disseminates garbage across the kitchen floor. At t_7 someone comes home and sees both the garbage and Annie at t_7. Should Annie at t_7 be held responsible for the actions of Annie at t_1? Well, one might think that if Annie at t_7 is a different object from Annie at t_1, as the perdurantist does (she thinks these are two numerically distinct objects), then we shouldn't hold one responsible for the actions of the other. Likewise, we might think that if Annie at t_1 is planning her day, and wondering what she ought to do later in the day, her deliberation only makes sense on the assumption that it will be her doing the things she now decides to do. What point would there be in deliberating about what to do if one is not around to do anything, and instead, someone else is around?

We will consider more fully the costs and benefits of endurantism and perdurantism later in this chapter. But before we do that, we need to spell out the views more rigorously. While there have been rigorous definitions of perdurance, it has proved less easy to define endurance. For what does it mean to say that something is *wholly present* at a time? Initially it might seem intuitive to say that to be wholly present at a time is to have *all* of your parts present at that time. After all, what is definitive of perduring objects is that they are only partially present at each time: they have parts (temporal parts) present at different times. We want to say that all of an enduring object is present at each time it exists, and surely, since the object is composed of parts, that must mean that all of its parts are present at that time. But of course, even the endurantist will allow that enduring objects gain and lose spatial parts over time. Annie is gaining and losing atoms every day, even if she does not ever lose a paw or a tail. But then there are times at which Annie has parts that she does not have at other times. So it cannot be that Annie has *all* of her parts at every time she exists. So perhaps we should say that at t, Annie has all of her t-parts present. But that won't distinguish endurance from perdurance, since even the perdurantist will agree that at t, Annie has all of her t-parts present.

If one appeals only to the notion of parts, and which things have which parts, then it is very difficult to say exactly what it is for an object to endure. Fortunately, more recently, it has become possible to spell out these competing views in terms of *location relations*. These are the relations that objects bear to the regions of spacetime at which they are located. In section 7.3 we will have more to say about these relations, and about how to define a number of views about persistence in terms of them. First, however, we want to pause briefly to consider the distinction between relationalism, substantivalism and supersubstantivalism about spacetime, since the difference between these three views will inform our subsequent discussion of the difference between endurantism and perdurantism.

7.2. Relationalism, Substantivalism and Supersubstantivalism

Suppose Sara asks you where you are. You will probably answer her by telling her where you are relative to some other things (such as, you are near the Law School, opposite the muffin stand). According to one view about the nature of spacetime, in some good sense this is all there is to say. *Relationalists* about spacetime hold that what exists are a bunch of spatio-temporal relations that obtain between objects or properties. Sometimes this is framed as the claim that for relationalists, spacetime is not real, and sometimes it is framed as the claim that for relationalists, spacetime is nothing more than the existence of these spatiotemporal relations, and the things that stand in these relations. So relationalism is dualistic in the following sense: it posits (at least) two fundamental kinds of things: spatio-temporal relations, and the things that stand in those relations (usually thought to be objects).

We can contrast relationalism with substantivalism. Substantivalism is the view that spacetime is an independent substance: it is not simply to be identified with the existence of spatiotemporal relations and the things that stand in those relations. Dualistic substantivalism is the view that spacetime exists, and is distinct from the spatiotemporal relations that obtain between objects and properties.

Most substantivalists will say that the spatiotemporal distances between objects *depend* on the spatiotemporal distances between the regions at which those objects are located. So, like the dualist relationalist, the dualistic substantivalist thinks there are (at least) two fundamental kinds of things. It is just that the two views disagree about what those two kinds of things are. Both agree that objects exist and are fundamental, but where the dualistic relationalist thinks that spatiotemporal relations are funda-mental, the dualistic substantivalist thinks that spacetime is fundamental. It is worth noting, at this point, that substantivalists do not need to accept anything like the view that spacetime is like an empty container into which we can put things, such that spacetime remains the same regardless of what is in it. This might be a natural view to have, but it sits uneasily with the general theory of relativity, which says that objects and properties in spacetime alter the shape of the spacetime they are in. Gravity, as it turns out, is the deformation of spacetime by massive objects. So it is consistent with substantivalism that though spacetime is distinct from objects, nevertheless the properties that spacetime has are in part determined by the objects that are in that spacetime.

Monistic substantivalism, also known as supersubstantivalism, is the view that spacetime is an independent substance, and that objects are identical with regions of spacetime. So, unlike the dualistic substantivalist, who thinks that objects and spacetime are two distinct fundamental kinds of thing, the supersubstantivalist thinks that objects just are regions of spacetime. This doesn't mean that supersubstantivalists hold that every region of spacetime is identical with some object (though they might think this); it just commits them to thinking that in so far as there are any objects, each object is identical with a region of spacetime.

It is probably true to say that at the present time substantivalism is more widely accepted than relationalism, in part because it is generally thought that substantivalism fits better with the theory of general relativity. There are far too many arguments for and against substantivalism and relationalism to consider them all here. Instead, in what follows we will assume that dualistic substantivalism is true, and then spell out four different views about persistence in terms of the different relations objects bear to regions of spacetime.

7.2.1. Substantivalism and Location Relations

The dualistic substantivalist holds that objects are located at regions of spacetime. So she is going to want an account of the relations objects bear to spacetime regions. We usually call these relations *location relations*. To begin, we need to take some location relation as primitive, and define the others in terms of that relation. We will take the primitive location relation to be *exact location*. Though there is no definition of exact location (because it is a primitive) the rough idea is that an entity, X, is exactly located at a region, R, if and only if X has (or has-at-R) exactly the same shape and size as R, and stands (or stands-at-R) in all the same spatial or spatiotemporal relations to other entities as does R. The idea is that an object's exact location is like its 'shadow' in spacetime. It is the region of spacetime that has exactly the same shape and size as does the object itself.

We can then define the following additional location relations in terms of exact location:

> **X is contained in R** iff X is exactly located at a proper or improper sub-region of R.

So X is contained in R if and only if none of X is *outside* of R. So for instance, if Annie is sitting in the kitchen then she is contained in the kitchen, even though much of the kitchen is free from Annie. Annie is contained in the kitchen because none of Annie is outside the kitchen.

> **X fills R** iff each sub-region of R overlaps an exact location of X.

So X fills R if and only if no sub-regions of R are free from X. So, for instance, consider Annie in the kitchen, and consider the region where one would say her foot is located. Annie fills that region, since none of that region is free from Annie. But there are regions disjoint from (entirely separate from) that foot-shaped region that arc also not free from Annie: the head-shaped region, for instance. So Annie can fill a region without being contained in that region. Annie fills the foot-shaped region, but she is not contained in it, because, in effect, she 'spills outside' of that region: she fills that region, and then some!

> **X is wholly located at R** iff X both fills and is contained in R.

X is wholly located at R if and only if not only does X fill R, but is also contained in R. That is, there is nowhere inside R where we fail to find X, and there is nowhere we find X, that is outside of R. If we are looking for X, we should look just in R: for there is nowhere in R we *won't* find X, and there is nowhere outside of R that we *will* find X.

In what follows we will use these location relations to define and explore four different accounts of persistence.

7.2.2. Three Accounts of Persistence

Persisting objects are objects that exist at different times. Since most ordinary objects have three dimensions at any given time (they have length, breadth and height), most ordinary objects occupy a three-dimensional region of space at each time they exist (the region of space that is the exact same size and shape as the object at that moment in time). So most ordinary objects occupy a sequence of three-dimensional regions: a different three-dimensional region at each time. Different theories of persistence crucially disagree about what relation a persisting object bears to each of these three-dimensional regions it occupies. Endurantists think that all of the object is located at each of these three-dimensional regions, while perdurantists think that only a part of the object is located at each of these regions.

The difference between different theories of persistence can be made clearest, and most stark, if we suppose (as presentists do not) that spacetime is four-dimensional (see Chapter 4 for further discussion of this view). If spacetime is four-dimensional then persisting objects will not only be located at three-dimensional regions of space, but also at some four-dimensional region of spacetime: namely the region that is made up of the sequence of three-dimensional regions the object is located at. In effect, we can think of this four-dimensional region as a persisting object's *path* through spacetime. It's a view of the whole object throughout its life.

It seems clear that any persisting object is wholly located at such a four-dimensional region. If a four-dimensional region is a persisting object's path through spacetime, then that region must *contain* the persisting object (the object can't be anywhere outside that region, otherwise the region wouldn't be its path) and the object must *fill* that region (otherwise some of the region wouldn't be its path at all).

So, we have agreement that persisting objects are located at a series of three-dimensional regions, and that persisting objects are wholly located at some four-dimensional region. The disagreement lies in *which* location relation persisting objects bear to each of the three-dimensional regions at which they are located. We can call each of these three-dimensional regions an M-region. Each M-region of some persisting object is a region that contains, and is filled by, that object *at a moment of time*. So the M-region of Annie, as she sits in the kitchen, is the dog-shaped region at which we find Annie. Now we have the tools to get to the nitty gritty of the different views of persistence.

According to endurantists, persisting objects are *exactly located* at each M-region, and are *not* exactly located at the four-dimensional region at which they are wholly located. The idea is that enduring objects are wholly located at a four-dimensional region *by being exactly located at a series of M-regions*, one for each time at which the object exists. This is meant to capture the sense in which enduring objects are wholly present at each moment. We can define endurance as follows:

X endures = *df* X is wholly located at a four-dimensional region, R, and X is exactly located at each M-region of R.

Hence if X endures, it persists by being *multi-located:* by being exactly located at a number of distinct M-regions. As we might say informally, all of X is located at some M-region, and all of X is located at the next M-region, and so on for each M-region: hence X is multiply located (see Figure 15).

The perdurantist agrees with the endurantist that there is *something* exactly located at each M-region of the four-dimensional region at which a persisting object is wholly located. But she doesn't think that the thing that is exactly located at each of those M-regions is X itself. She thinks that what is exactly located at each M-region is an instantaneous temporal part of X. X *itself* is the sum of each of those distinct temporal parts. Or, as we might say, X is *composed of* all of those temporal parts. Given this, we can define perdurance as follows:

X perdures = *df* (a) X is wholly located at a four-dimensional region, R, and (b) X is exactly located at R, and (c) for any M-region of R there exists an X* such that (i) X* is exactly located at R* and (ii) X* is a proper part of X.

So perdurantism is the view that, at each time, a *distinct* three-dimensional object exists (the thing that is exactly located at the relevant M-region) and the persisting object is the thing composed of each of these distinct three-dimensional objects (see Figure 16).

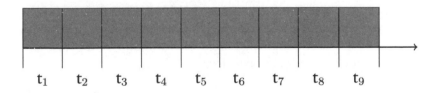

Figure 15 Endurantism: The entire object is wholly located at each moment that it exists. The enduring object, X, is exactly located in each of the squares, and is wholly located in the totality of the long rectangle. (That each moment appears to have temporal width is an artefact of the diagram.)

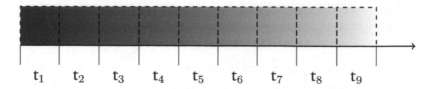

Figure 16 Perdurantism: The object is not wholly located at each moment that it exists. Rather, the object is partially located at each moment that it exists. Each temporal part of X is exactly located in each of the squares, and X is wholly located at the entire rectangle. X is also exactly located at the entire rectangle, because X is the sum of all of the 'square' temporal parts (again, it is an artefact of the diagram that each temporal instant appears to have temporal width).

We can now see that a third option presents itself. For one can imagine someone who accepts, as the perdurantist does, that there is something that is wholly and exactly located at some four-dimensional region, R, but who rejects the claim that *anything* is exactly located at any M-region of R. We can call this view *transdurance*. A transduring object is located at each M-region of R by filling each of those regions; but transduring objects fail to be contained in any such region because some of the transduring object spills outside of any M-region.

One way to think about transdurance is as the view that persisting objects are stretched along the temporal dimension (as perdurantists suppose) rather than being multi-located along it (as endurantists suppose). But they are stretched along the temporal dimension without having parts at each M-region. They are mereologically simple along the temporal dimension (mereologically simple just means lacking parts). Since they are also extended along that dimension, they are (along that dimension, at least) extended simples.

So, when we meet a transduring object at a time, we are not meeting all of the object (since some of it is elsewhere) as we would be meeting all of an enduring object; nor are we meeting only a proper part of the object (as we would a perduring object) since it has no proper part present at the time. We are, as it were, meeting *some* of it, without meeting any part of it. We can define transdurance as follows:

X transdures = *df* (a) X is wholly located at a four-dimensional region, R, and (b) X is exactly located at R, and (c) either (i) there is no object, X*, that is exactly located at any M-region of R, or (ii) any object, X*, that is exactly located at any M-region of X is not a part of X.

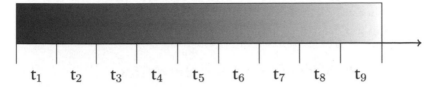

Figure 17 Transdurantism: The transduring object is wholly located at a four-dimensional region of space-time (R), which is the entire shaded rectangle. It is also exactly located at that same region. The object is simple and so lacks temporal parts. Accordingly, the object is not partially located at any moment at which it exists. (Again, the appearance as of each time itself being extended is an artefact of the diagram.)

This definition is just a complicated way of saying that a transduring object is exactly located where it is wholly located, and that either nothing is exactly located at any of the M-regions of R, or if there is something thus located, it is not a part of the transduring object. Perhaps, for instance, there is some object that exactly occupies an M-region of R, but it doesn't share any parts with X (see Figure 17).

Notably, in all three diagrams we see that the persisting object is located at the very same 'rectangle' (i.e. the same four-dimensional region): what differs is the way it is located at each of the three-dimensional sub-regions. That is the crucial difference between different theories of persistence. Of course, if there is no four-dimensional spacetime (as presentists hold), then all that exists is a single three-dimensional region. But since there exists a *sequence* of such regions, we can still understand the competing views in much the same way. However, presentism is more naturally paired with endurantism, since if we pair it with perdurantism or transdurantism we have to say that most of any persisting object doesn't exist, and while that is not incoherent, it is not very attractive.

7.3. Evaluating the Accounts

In what follows we will consider just a few of the biggest obstacles to each of the different accounts of persistence, and the responses to those obstacles on the part of defenders of those views. Our aim is to provide a sense of some of the problems the views face, rather than providing an exhaustive explication and consideration of objections and responses. Note that we will simply assume that all four accounts of persistence are candidates to describe the manner in which actual objects persist. In the final section of the chapter we move on to consider whether all four views of persistence are consistent with all of the views about temporal ontology that we met in Chapter 1.

7.3.1. The Problem of Temporary Intrinsics

A perfectly general problem arises when we consider the nature of persisting objects. This problem is known as the problem of temporary intrinsics. The 'temporary' here refers to the fact that persisting objects seem to change over time, and hence to instantiate *temporary* properties. 'Intrinsics' refers to intrinsic properties. Intrinsic properties are properties that an object has in virtue of how that object is, in and of itself, rather than in virtue of the relation that the object bears to other objects. So consider one property Sally has: the property of having a white car. This is clearly not an intrinsic property. It's a property she has in virtue of bearing a relation to something else. Intuitively, she could sell the car and nothing about how she is, in herself, would change. By contrast, her mass, height, colour, bone density, what she is made of, and so on, are all intrinsic properties of Sally. If Sally's mass, height, colour, or what she is made of change, then she changes her intrinsic properties. Since all of these properties do, in fact, change (even if only slightly) Sally has a range of temporary intrinsic properties. The problem of how to accommodate temporary intrinsic properties is known as the problem of temporary intrinsics.

We can divide the problem of temporary intrinsics into two sub-problems: the problem of change and the problem of incompatible properties. The problem of change is the problem of reconciling the indiscernibility of identicals with the change over time in persisting objects. The indiscernibility of identicals says that X = Y only if X and Y share all of the same properties. It is easy to see why that is a plausible principle. If X and Y are one and the same thing, it is hard to see how one could have a property that the other lacks. If we found out, for instance, that Clark Kent has a property that Superman lacks (say, Clark Kent cannot fly) we would surely be right to conclude that Clark Kent is not identical to Superman. There is not one guy that we pick out with two names; there are two guys! Still, in some good sense what it is for O to persist is for O to exist at more than one time. Yet it is unclear how one and the same object *can* exist at multiple times, and, consistent with the indiscernibility of identicals, instantiate different properties at those times. Hence it is not clear how persisting objects *can* change if the indiscernibility of identicals is true.

The second, related, problem is the problem of incompatible properties. Suppose that Annie is tired at one time, and not tired at another time. Then Annie is both tired and not tired. But nothing can be both tired and not tired, for that would be to have incompatible properties. So either Annie has incompatible properties, or it is not the case that she is tired at one time, and not tired at another, and hence not the case that she instantiates temporary intrinsic properties. Jointly, the problem of change and the problem of incompatible properties make it difficult to see how persisting objects can instantiate temporary intrinsic properties: hence the problem of temporary intrinsics.

7.3.2. Perdurance and Temporary Intrinsics

Perdurantists have a fairly straightforward solution to the problem of temporary intrinsics. Since perdurantists hold that the way objects are extended along the temporal dimension is the same as the way they are extended along the spatial dimension, they can offer the same account of the way objects change over time as they offer about how objects instantiate different properties across space. Consider Annie. At any given time she is black at some regions, and dark chocolate at other regions (notably her beard). So she instantiates different properties *across space*. She does this by having different spatial parts: she has black parts and chocolate parts. So we can say that Annie 'changes' across space by having different parts with different properties. Each part is just black, or just chocolate – the parts themselves do not change – but Annie 'changes' by having both black objects and chocolate objects as parts. The perdurantist will say the same thing about how objects change *over time*. Annie is tired at one time, and not tired at another, by having a tired part (a tired temporal part) and a not tired part (a not tired temporal part). So Annie herself changes by being composed of (being made up of) a series of distinct, unchanging, instantaneous objects (her temporal parts) each of which instantiate different properties. The problem of change dissolves, on this view, since each of the objects that instantiates different properties (tiredness, on the one hand, and non-tiredness, on the other) are numerically distinct objects, and hence the indiscernibility of identicals does not apply to these objects. The whole object, Annie, instantiates a single, unchanging, set of properties. But these properties are properties such as having a tired part and having a non-tired part, *both* of which Annie instantiates. Yet she can instantiate both these properties and change over time! Problem solved. This account also provides a solution to the problem of incompatible properties. For Annie *is* both tired and not tired: but she is tired by having a tired temporal part, and not tired by having a non-tired temporal part. But there is nothing contradictory in this; just as Annie can be both black and chocolate by having a black part and a chocolate part – it is just that she cannot be both black all over and chocolate all over – so too she can be both tired and not tired, so long as she is tired at some times, and not at others, by virtue of having tired (and not tired) temporal parts.

7.3.3. Endurance and Temporary Intrinsics

The view that persisting objects endure is the view that persisting objects are multiply located. It is the view that Annie, exactly located at one time, is numerically identical with Annie, exactly located at any other time. Endurantists who are not also presentists straightforwardly face the problem of temporary intrinsics since, unlike the perdurantist, they think that the three-dimensional object that is Annie at one time *is* numerically identical with the three-dimensional object that is Annie at some other time, despite the fact that at each of those times Annie instantiates different properties. Presentists can avoid this problem since they think that only one time exists. So what is true of Annie is what is true of her in the present: there is no other way she is, in the past (or future) and hence no contradiction arises.

Still, endurantists can respond to this problem without endorsing presentism. They can note that, unlike the perdurantist, they treat extension through time very differently from extension through space. Given this, endurantists suggest that the properties persisting objects have at times are really properties they have *relative* to times, or properties they have in a temporally relativised *manner*. So suppose that at t_1, Annie has the property of being a puppy, and at t_{20}, she has the property of being an adult dog. Then she has temporary intrinsic properties (or so it seems). The endurantist will suppose that there are two different ways one can understand this case (and others like it). According to the first, Annie doesn't really have intrinsic properties. Instead, instantiating any property involves bearing a relation to a time. So really, Annie bears the relation of being-a-puppy-at to t_1. She also bears the relation of being-an-adult-dog-at to t_{20}. There are no intrinsic properties, just relations objects bear to times. This solves both the problem of change and the problem of incompatible properties. The endurantist can say that Annie has a single set of these relativised properties. Annie bears *both* the being-a-puppy-at and the being-an-adult-dog-at relations, it is just that she bears them to two different times. That allows the endurantist to make sense of how Annie can both change and yet it be the case that there is some single set of properties and relations that she bears. It also allows the endurantist to make sense of the apparently incompatible properties that Annie instantiates. For there is, in fact, nothing incompatible in Annie bearing both the being-a-puppy-at and the being-an-adult-dog-at relations, so long as she bears those relations to different times.

According to the second strategy endorsed by endurantists, rather than saying that Annie has no intrinsic properties, and that what appear to be properties are really disguised relations to time, instead the endurantist holds that Annie instantiates the properties she does in a certain *temporal manner*. To get a handle on this, let's consider an analogy. Suppose that Annie is running. She might be running *quickly*, or running *slowly*, or running *sideways*: these are all *ways* that Annie runs. According to the endurantist strategy under consideration, then, Annie always instantiates properties in a particular temporal way, just as she always instantiates running in some particular way or other (fast, slow, etc.). So if Annie has the property of being a puppy at t_1, then we should understand this to mean that she instantiates the property of being a puppy in a $t_1$ly manner. By parity, Annie instantiates the property of being an adult dog in a t_{20}ly manner. As before, this strategy allows the endurantist to resolve both the problem of change and the problem of incompatible properties. There is a single set of properties that Annie instantiates, a set of properties that includes being a puppy *and* being an adult dog. It is just that the former property is instantiated $t_1$ly, and the latter t_{20}ly. This allows the endurantist to reconcile change with the indiscernibility of identicals. The same strategy also solves the problem of incompatible properties, since there is nothing incompatible in Annie instantiating by both being a puppy and being an adult dog, so long as the temporal manner of instantiation of each of these properties is different (which it is).

7.3.4. *Transdurance and Temporary Intrinsics*

Transdurantists, too, face the problem of temporary intrinsics. Suppose that a transduring object starts off life being short, and later in life is tall. How do transduring objects instantiate different properties at different times without falling foul of the indiscernibility of identicals? And how do they do so without instantiating incompatible properties? The transdurantist can appeal to the same sort of machinery to which the endurantist appeals. She can say either that transduring objects do not instantiate intrinsic properties, but instead bear relations such as being-tall-at and being-short-at to times. Or she can say that transduring objects instantiate properties in a particular temporal manner (tly). In either case, she will say that the whole transduring object instantiates a single set of compatible properties. The properties are compatible because an object can be both tall and short, so long as it bears the relations of being tall and being short to different times, or instantiates those properties in a different temporal manner.

7.3.5. Perdurance and Identity

We have seen how perdurantists can, relatively straightforwardly, avoid the problem of temporary intrinsics. Still, it is the very mechanism by which perdurantists avoid the problem of temporary intrinsics that creates new problems, and it is these problems we will now consider.

The objection most commonly levelled at perdurantism is that it is not really an account of persistence proper at all, since it fails to make sense of important features of persisting objects. After all, according to this view what exists is a series of numerically distinct temporally unextended objects and the sums of those objects. But think about objects like ourselves, who remember, and deliberate, and plan, and regret and anticipate, and are held responsible for past actions. It has been objected that if all that exists is a series of numerically distinct things and the sum of those things, then we cannot make any sense of these practices. Since these practices clearly do make sense, this is reason to be sceptical of this view of persistence.

Endurantism, by contrast, can clearly accommodate the idea that one ought sometimes to regret past actions, to be held responsible for such actions, to anticipate future experiences, and so on. That is because the person in the past who performed those actions is numerically identical with the person doing the regretting now. Similarly, the (relevant) person in the future is numerically identical with the person now, which makes it rational to anticipate that person's experiences, and to plan for that person's future. But none of that is so if perdurantism is true. When you meet Sally today, and then again tomorrow, you meet two distinct persons; indeed you meet two persons who are as distinct from one another as you are distinct from Sally. Yet it makes no sense for you to anticipate Sally's experiences, or feel bad for Sally's actions. So how can it make sense for Sally today to anticipate the experiences of Sally tomorrow, when Sally today is numerically distinct from Sally tomorrow?

The perdurantist will respond to this worry by noting that although it is true that the three-dimensional person-like object that exists today is *distinct* from the three-dimensional person-like object that exists tomorrow, it is not true that the relation that obtains between those two things is the same as that which obtains between you and Sally today. After all, consider the following question: what makes some sum of dog temporal parts a *dog*? On many views of composition there is something that is composed of a poodle temporal part today, and a collie temporal part tomorrow, and a pug temporal part the day after. The perdurantist will say (if she accepts our earlier definition) that such a thing perdures. This might seem counterintuitive. The way to make this less counterintuitive is to notice that she will not say that that four-dimensional object is any ordinary kind of object. In particular, it is most certainly not a dog, despite the fact that it is composed of dog temporal parts. That's because what is required for something to be a dog is for its temporal parts to be connected in important ways. For a start, it matters that the temporal parts of a dog are causally connected in appropriate ways. It matters that the properties of the puppy temporal parts are the *cause* of the properties of the later adult temporal parts. That's because it matters that what happens to earlier temporal parts leaves a causal mark on later parts: that's why dog training matters – because what you do with earlier temporal parts affects later temporal parts. The difference between Annie and some random sum of dog temporal parts is that her parts are causally connected in an appropriate manner, and that is why she is a dog.

So there is a perfectly good sense in which the relation that obtains between Sally today and Sally tomorrow is different from the relation that obtains between you today and Sally today (or you today and Sally tomorrow), even though all of these objects are numerically distinct. That is because the temporal parts of Sally today and Sally tomorrow are causally connected in a way that you are not connected with Sally today or Sally tomorrow. That means that Sally tomorrow will bear various psychological relations with Sally today: it will remember many of the things that Sally today did, it will have many of the psychological characteristics of Sally today, and so on. None of that is true of you today and Sally today or tomorrow. So the perdurantist will appeal to the different sorts of connections that obtain between three-dimensional objects to explain why in some cases it is rational for one such object to anticipate the experiences of another, and in other cases it is not. More generally, they will appeal to these kinds of connections to explain why some sets of these objects are temporal parts or temporal counterparts of an ordinary object, and why others are not.

7.4. Persistence and Temporal Ontology

That brings us to our final section of the chapter. So far we have outlined three rival views about persistence, and briefly considered a few objections to these views and the responses to those objections. We have already seen that certain monistic views, according to which spacetime and objects are not distinct kinds of things, are inconsistent with certain views of persistence (endurance and transdurance) on the assumption that spacetime is four-dimensional. The question that now arises is whether each of these views is consistent with all views about temporal ontology, or whether some views about temporal ontology commit one to certain views about persistence.

It seems clear that all four views are consistent with an eternalist picture of time. What of presentism? Some have thought that endurantism and presentism are a natural mix. For notice (as we noted previously) that one way to solve the problem of temporary intrinsics is to think that only the present moment is real, and hence an enduring object only has the properties it has *now* (hence there can be no problem of incompatible properties). Equally, many have thought that perdurantism and trans-durantism are incompatible with presentism. The thought is that if only the present moment exists (or if only present entities exist), then both the perdurantist and the transdurantist are committed to holding that most of a persisting object does not exist. That is why perdurantists and transdurantists are typically eternalists. Still, the views are not obviously *incompatible* with presentism: one could potentially make sense of the idea that only *some* of an object ever exists, and there have been attempts to do just this.

If one can make sense of all four views given both eternalism and presentism, then it should be that they are also compatible with both the growing block view and the moving spotlight view, and that is what we find. So nothing about one's temporal ontology forces one to accept a particular view of persistence, though some combinations of views are rather more natural than others.

7.5. Summary

In this chapter we have looked in detail at the various theories of persistence that are currently available. Theories of persistence aim to help us understand what it is for something to exist through time. We have covered some quite technical material in this chapter. So let us take a step back and review the key points.

(1) There are, broadly, three theories of persistence available: endurantism, perdurantism and transdurantism.

(2) Underlying the distinctions between endurantism, perdurantism and transdurantism is a cross-cutting distinction between relationalist and substantivalist accounts of spacetime, and dualistic and monistic varieties of relationalism and substantivalism respectively.

(3) We can differentiate endurantism, perdurantism and transdurantism by defining these views in terms of location relations.

(4) If we appeal to location relations as the basis for differentiating endurantism, perdurantism and transdurantism then some version of dualistic substantivalism seems to be required.

(5) Certain versions of monistic substantivalism seem to rule out certain views of persistence, at least if those views are understood in terms of location relations.

(6) All theories of persistence need to account for temporary intrinsics, but they do this in different ways.

(7) The endurantist has two standard solutions to the problem of temporary intrinsics available to her. First, she can relativise the having of properties to times. Second, she can allow time to modify the way in which properties are had.

(8) The central problem facing perdurantism is the problem of explaining how we can make sense of deliberation, responsibility, blame and so on, since, on that view, it is not the case that numerically the same person exists from moment to moment.

7.6. Exercises

i. Consider each of the four theories of persistence outlined in this chapter. Identify any difficulties that you can see for each theory. These can be problems that we have identified, or problems of your own devising. Now identify any benefits of these theories that you can think of. Arrange all of this information in a table to try to get a sense of what the best theory of persistence might be.

ii. Consider the two solutions to the problem of temporary intrinsics that the endurantist has outlined. Can you see any problems for these solutions?

iii. Recall each of the views of temporal ontology from Chapter 1.

Consider the extent to which these views can be rendered compatible with each of the theories of persistence discussed here. Pair each view of temporal ontology with the most natural theory of persistence that is compatible with that view.

iv. In this chapter we have defined endurantism and perdurantism in terms of location relations. Do you think that this is the only way to define these views? Can you think of any other ways that you might define endurantism or perdurantism?

v. Consider the idea that there are multiple possible worlds. Can any of the views of persistence discussed here be transformed into views about what it might mean for an individual, Sara, to exist in multiple distinct possibilities?

vi. Discuss the view that nothing persists at all. Is there anything that might be said in favour of such a view?

vii. Remember the discussion of hypertime from Chapter 2. Can you develop a two-dimensional version of any of the theories of persistence discussed here? What advantages might such a theory have (if any)?

7.7. Glossary of Terms

Annie
Kristie Miller's black labradoodle.

Extended Simple
A mereologically simple thing that is extended in time, space or both.

Indiscernibility of Identicals
The principle according to which if X and Y are identical, then X and Y share all of the same properties.

Mereological Simple
Something is mereologically simple when it has no proper parts.

Numerical Identity
The relation of being the self-same object; a relation that every entity bears to itself.

Temporary Intrinsic
An intrinsic property that an entity has at some times but not at others.

7.8. Further Readings

S. Dasgupta (2015) 'Substantivalism vs Relationalism About Space in Classical Physics', *Philosophy Compass* 10 (9): 601–24. While this is not an introductory work, it is an accessible overview of the debate between substantivalism and relationalism about spacetime.

C. Gilmore, D. Costa and C. Calosi (2016) 'Relativity and Three Four-dimensionalisms', *Philosophy Compass* 11 (2): 102–20. While also not an introductory work, this is an accessible overview of the different theories of persistence.

R. Wasserman (2010) 'Teaching & Learning Guide For: The Problem of Change', *Philosophy Compass* 5 (3): 283–6. This is an introductory overview of the problem of change and its connection with theories of persistence.

8

The Paradoxes of Time Travel

Time travel stories have become one of the staple narratives of science fiction. In the twentieth century, the idea that we could travel in time became a real possibility with the advent of the special and general theories of relativity. As soon as scientists began to treat time travel as a live possibility, a number of critics voiced their concerns over the coherence of travel into the past. Travel into the past, it was suggested, would inevitably lead to paradoxes that threaten to 'tear apart the spacetime continuum' or force us into some otherwise horrible mess. This view of time travel was voiced by the famous science-fiction author Isaac Asimov, who once wrote:

> The dead giveaway that true time-travel is flatly impossible arises from the well-known 'paradoxes' it entails. The classic example is 'What if you go back into the past and kill your grandfather when he was still a little boy?' … So complex and hopeless are the paradoxes … that the easiest way out of the irrational chaos that results is to suppose that true time-travel is, and forever will be, impossible. (Isaac Asimov, *Gold: The Final Science Fiction Collection*, Harper Voyager, 2003, pp. 276–7)

In this chapter we will consider the case for Asimov's position. Is time travel really impossible? Does travel into the past inevitably result in a world torn apart? The answer is straightforward: no. But of course, as we shall see, the devil is in the details.

8.1. What is Time Travel?

Before we can look at the paradoxes of time travel in earnest, we need to first consider what time travel is. We want an account that specifies under what conditions someone or something travels in time: we want the necessary and sufficient conditions for some instance of travel to be *time* travel. That is to say, borrowing some terminology from Chapter 1, we want a constitutive theory of time travel. That's because we are not asking what actual time travel is like (if there is any) – whether it occurs through a wormhole, or via a machine like the TARDIS, or in some other manner. We are only asking what it would take for something to count as being time travel. With a better sense of the nature of time travel in hand, we can then turn to the potential for catastrophe that it supposedly entails.

8.1.1. *External Time and Personal Time*

We know what it is to travel through space. We do it every day when we head to the café to get coffee. We leave one location in space and travel to some other location. Our travel takes time. We can calculate the speed by which we move by seeing how much time elapses, per amount of spatial distance we cover. But what does it mean to talk about travelling in *time*? In the broadest sense, travel through time involves leaving one temporal location and arriving at some other temporal location. According to this broad definition, we are all travelling through time. Each of us leaves one temporal location and arrives at some other temporal location. Each of us left yesterday, and arrived at today, and consequently we time travelled from yesterday to today. That, however, is not the kind of time travel in which philosophers are interested. Indeed, that sense of time travel is better known as *persistence*, a phenomenon we considered in the previous chapter. In what follows, when we talk of time travel we will mean time travel other than ordinary persistence. Call this *interesting* time travel.

Lewis's seminal 1976 paper 'The Paradoxes of Time Travel' offers a useful way both to conceptualise interesting time travel and to mark the difference between interesting time travel and mere persistence. Lewis introduces a distinction between *external time* and *personal time*. External time is just time. The external temporal distance between two events is the temporal distance between those events. So the temporal distance between December 2015 and December 1985 is thirty years of external time.

Personal time, by contrast, is not really *time* at all. Personal time is really a measure of change. While the traveller is travelling, he gains new memories, gets hungry, eats, digests food, his fingernails grow, his cells die, and so on. These changes typically take a certain amount of external time. For instance, it typically takes about one month of external time for fingernails to grow 3mm. So suppose Fred, the time traveller, gets into a time machine in January 2017, and while in the time machine his nails grow 3mm. If all of the rest of the changes Fred undergoes are roughly those you and I would undergo through a period of a month (if he accrues a month's worth of memories, eats a month's worth of dinners, and so on) then we will say that one month of personal time has elapsed for Fred. If, at the end of his journey, Fred gets out of the time machine and is in January 2000, then it has taken him one month (of personal time) to travel seventeen years (of external time).

According to Lewis, a journey counts as interesting time travel only if there is a disparity between personal time and external time. Notice that in the case of persistence there is no disparity between personal time and external time: it takes each of us one day of personal time to travel one day of external time.

> **Personal Time:** n minutes of personal time elapses, iff the amount of change that occurs is the same as the amount of change that would typically occur in n minutes of external time.

Lewis held that the disparity between personal time and external time is a necessary condition for a journey to count as (interesting) time travel. We can put this as follows:

> **Time Travel Necessary Condition 1:** Necessarily, a journey counts as time travel only if there is a disparity between the elapsed personal time of the traveller during that journey, and the external time travelled during that journey.

There are two ways for personal time and external time to come apart. First, the amount of personal time that elapses might be different from the amount of external time that elapses. Second, the temporal relations between events in personal time might come apart from the temporal relations between events in external time, without the amount of time actually differing. This second way for personal time and external time to differ might be difficult to understand. So here's an example to help. Suppose that Fred gets into a time machine at 2pm on 2nd January 2015, and steps out of the time machine on 1st January 2015. In terms of external time, he has travelled one day into the past. Suppose, however, that his journey in personal time itself takes a day. In this case there is no disparity between Fred's personal time and external time with respect to the amount of time that has passed. There is, however, a disparity with respect to the temporal relations between the start of Fred's journey and the end of his journey in the two times. In external time, the start of Fred's journey occurs after the end of his journey. In personal time, the start of Fred's journey occurs before the end of his journey. This helps to bring out one of the important features of personal time: personal time always ticks forwards *even when* external time ticks backwards.

In cases of persistence there is no disparity of either kind. If a time traveller sits in her office and waits for an hour, then the start of her journey and the end of her journey are an hour apart in external time. But so too for her personal time: she has to wait for an hour. Moreover, the start of her journey is before the end of her journey in external time. Again, so too for her personal time: the start of her waiting period in personal time happens before the end of that period.

The distinction between personal time and external time therefore provides us with a way to differentiate persistence from interesting time travel. In cases of persistence, external time and personal time are in accord in the two ways discussed. In cases of time travel, external time and personal time are out of sync in one of the two senses described.

What else does Lewis think is necessary for a journey to count as time travel? Well, in order for anyone to count as travelling from place A to place B, it needs to be that the person who leaves A is the same person who arrives in place B. After all, Sara can't travel to Singapore from Sydney by staying in Sydney and having a friend of hers in Thailand travel to Singapore. Nor would she count as having travelled to Singapore if she was killed in Sydney, and, by pure fluke, someone just like her was created in Singapore. So a second necessary condition for a journey to count as time travel is that the person at the beginning of the journey is the same person as the person at the end of the journey. Indeed, since objects that are not persons can travel in time (we could send an urn back in time) any object, O, counts as travelling in time only if the object that departs on the journey, O, and the object that arrives at the end of the journey, O*, are one and the same object.

Time Travel Necessary Condition 2: Necessarily, a journey counts as time travel only if the traveller who departs and the traveller who arrives are one and the same traveller.

Lewis thinks that these two necessary conditions are, jointly, sufficient conditions for a journey to count as time travel. That is, he thinks if both of the necessary conditions are met, then the journey is an instance of time travel.

> **Time Travel Sufficient Condition:** Necessarily, a journey counts as time travel if (a) there is a disparity between the elapsed personal time of the traveller during that journey and the external time travelled during that journey, and (b) the traveller who departs and the traveller who arrives are one and the same traveller.

As we saw in the previous chapter, what it takes for an object at one time to be the *same* object as some object at another time is controversial. But, with the tools of Chapter 7 in hand, we can say that something counts as going on a journey only if the thing that leaves and the thing that arrives are the same persisting thing (leaving it open exactly what it takes for an object to persist).

8.1.2. Forwards Travel versus Backwards Travel

Backwards time travel is time travel towards the past. Forwards time travel is travel towards the future. Much of the discussion within philosophy has focused on backwards time travel. But forwards time travel is just as interesting. Indeed, according to Einstein's theories of general and special relativity forwards time travel is relatively straightforward.

In order to understand time travel in the context of special and general relativity, we need to complicate the distinction between personal time and external time somewhat. According to the special and general theories of relativity, the temporal distance between events depends upon one's relative state of motion. This has a range of important implications, some of which were discussed in earlier chapters. The crucial point is that this variance in temporal distance gives rise to an important relativistic phenomenon: the phenomenon of time-dilation. As we increase our speed, external time slows down. So, for instance, suppose that we place a clock on Earth and we place a clock on a spaceship that is travelling away from the Earth at half the speed of light (150,000 km/s). Suppose that the clock on the spaceship records that 15 minutes has passed. In the time it takes the clock on the spaceship to record the passage of 15 minutes, the clock on Earth will have recorded the passage of approximately 23 minutes. As we speed up, the discrepancy grows exponentially. This means that if one were to travel at very close to the speed of light in a spaceship travelling on a roundtrip away from the Earth, minutes or hours may have passed for the occupant of the spaceship, while thousands of years might have passed on Earth.

Because the temporal distance between events depends on one's relative motion, special and general relativity recommend multiplying external time. There is no longer just one external time. Rather, because the temporal ordering and temporal distance between events depends on relative motion, there are many external times, each of which is assigned to an inertial frame of reference: a perspective on the universe that is in constant motion. Accordingly, we can now differentiate between external time 1, external time 2, external time 3 and so on. Each observer still has only a single personal time. It is just that a single observer's personal time may correspond to different external times as they change their relative state of motion. It doesn't follow from this, however, that personal time and external time are out of sync through such shifts in relative motion. So long as the personal time of an observer within a frame of reference always corresponds to the external time within that frame of reference, then personal time and external time remain in sync. Which is to say that if one speeds up and external time slows down, then so long as one's personal time also slows down to match the external time, the two are kept in sync. And, indeed, that is precisely what happens.

This is just to say that persistence has its analogue in a relativistic setting. So too does the interesting type of time travel. The notion of interesting time travel identified above, however, needs to be modified somewhat. Recall the necessary condition that Lewis placed on time travel:

Time Travel Necessary Condition 1: Necessarily, a journey counts as time travel only if there is a disparity between the elapsed personal time of the traveller during that journey, and the external time travelled during that journey.

This definition is fine for pre-relativistic time, but needs to be updated for relativistic time. We continue defining time travel in terms of discrepancies between personal time and external time; however, in a relativistic context we must understand these discrepancies with respect to a particular external time. If one's personal time is out of sync with external time 1, then one has time travelled with respect to external time 1. It is compatible with one's personal time being out of sync with external time 1, however, that it remains in sync with some other external time, such as external time 2.

To grasp the idea, suppose that Bert and Ernie are identical twins living on Earth. Suppose we put Bert into a rocket that travels at high speed away from the Earth, reaches a distant destination, then turns and heads back towards Earth. When Bert arrives back on Earth and gets out of the rocket, he has aged six years, while Ernie has aged ten years. So the twins are no longer the same age! How can this be? Well, when Bert speeds away from the Earth his external time slows down relative to Ernie's personal time. Bert's personal time syncs with his new external time, and so this slows down too. That means Bert ages more slowly compared to Ernie. When Bert returns to Earth, his external time shifts back to Ernie's external time. His personal time syncs back up with Ernie's external time, and the two start ageing at the same rate once again: a year older for Ernie will be a year older for Bert, despite the new age difference between the two twins. With respect to Ernie's external time, Bert is a time traveller. Although Bert's personal time re-syncs with Ernie's once he arrives back on Earth and he starts ageing at the same rate, Bert's personal time remains somewhat out of sync with Ernie's external time in a global sense. According to Bert's personal time, only six years have passed when he arrives back on Earth. According to Ernie's external time ten years have passed. So there is a discrepancy between Bert's personal time and Ernie's external time.

It is worth noting that there is also a discrepancy between Ernie's personal time and Bert's external time. For Ernie, ten years have passed in his personal time. In Bert's external time, however, only six years have passed. Does this make Ernie a time traveller with respect to Bert's external time? According to Lewis's account of what it is for someone to be a time traveller, the answer is 'yes'. But you might think not. If you have the view that Ernie is not a time traveller with respect to Bert's external time, then you might try to modify Lewis's definition to handle this sort of case. We leave it to the reader to consider strategies for how this might be achieved.

Forwards time travel of the kind just described is relatively easy. In fact, it's been done: astronauts on the international space-station are travelling fast enough that they are shifted into a distinct external time relative to us here on Earth. Those astronauts are travelling micro-seconds into our future. Backwards time travel, by contrast, even in the context of the special and general theories of relativity, is much harder to achieve. This has a lot to do with the causal structure of the universe according to relativity. Travel into the past requires retro causation: the causation of events in the past by events in the future.

Retro causation, although compatible with general relativity, is quite difficult to achieve. Still, it can happen. If our world is correctly described by relativity then it is possible for there to be *closed time-like curves*. A closed time-like curve is the path of an object through spacetime where that object's path returns to its original position. One way in which an object can have a path that returns to its starting point is if spacetime itself is curved. To see this, pick up a piece of paper and turn it into a cylinder by connecting its two ends. Now trace a path around that cylinder. The path can begin at one location on the cylinder, and, despite always going in the 'same' local direction, will end up back where it started. We can imagine that a sub-region of spacetime is folded up in just this way and thus that time curves back on itself. Then an object that travels along this curved spacetime for some period will, for some portion of its lifetime, travel along a closed time-like curve. During that period the object will always seem to be moving forwards on a straight path, but by doing so it will travel back to an earlier external time. Since the existence of closed time-like curves is entailed by some solutions to Einstein's field equations, we know that such curves can exist. So far, however, no one has ever found a region of spacetime that is folded in this way (though such folding may occur near a black hole).

8.1.3. Possibility: Logical, Metaphysical and Nomological

As we noted, philosophers tend to focus on backwards time travel. This is because backwards time travel is thought to give rise to various unnerving paradoxes. In a relativistic setting, however, the alleged paradoxes of backwards time travel can be formulated just as easily for forwards time travel of the kind described. Philosophers tend to focus on backwards time travel, however, as it is easier to elucidate and discuss the kinds of paradoxes that are thought to arise. Moreover, in the backwards time travel case, these paradoxes will arise even for non-relativistic universes, which may not be true for the forwards-looking versions of the paradoxes. At any rate, let us follow tradition here and focus on the backwards case.

One of the questions philosophers ask about time travel is whether it is possible. As we have seen, possibility is a modal notion, which is often understood in terms of *possible worlds*. The idea is that something is possible if there is some world at which that thing occurs. Something is impossible if there is no world at which it occurs. So, for instance, it is impossible that there are square circles, because there is no world in which there are square circles. To capture this idea philosophers often say that square circles are *logically impossible*. That is because there is something contradictory about the very idea of a square circle, since a square circle would need to be both circular and not circular, and being both P and not P is a contradiction. So X is logically impossible if X's obtaining would result in a contradiction obtaining. Since contradictions are impossible – there are no worlds in which both P and not P – anything that entails a contradiction is thereby logically impossible.

Often when philosophers ask whether time travel is possible they are asking whether it is logically possible. They are asking whether there are any worlds in which there is time travel (backwards time travel, that is). That is the question we will consider in the next section.

At other times, philosophers ask whether time travel is *metaphysically possible*. Sometimes metaphysical possibility is taken to be the broadest kind of possibility there is. Understood as such, the sphere of the metaphysically possible worlds is the same as the sphere of the logically possible worlds – namely, all of the worlds. Sometimes metaphysical possibility is defined by the set of worlds that share the same metaphysical truths as our world (i.e. truths about the nature of reality; truths that are captured by or enshrined within our best metaphysical theories of the world). This is what we will mean by metaphysical possibility. If one thinks that all worlds share the same metaphysical truths, then one thinks that the sphere of the metaphysically possible worlds is the same as that of the logically possible worlds. But if one thinks that some worlds have different metaphysical truths from our world, then one thinks that some worlds are logically possible, but not metaphysically possible. Hence philosophers who ask about the metaphysical possibility of time travel are asking whether the *metaphysical* truths are consistent with time travel.

Finally, philosophers sometimes ask whether time travel is *nomologically possible*. Then they want to know whether time travel is consistent with the laws of nature; that is, whether, out of all of the possible worlds that share the same laws of nature as our world, any of those worlds contain time travel.

As we have already seen, as far as we know, time travel is permitted by the special and general theories of relativity. Of course, it doesn't follow that time travel *is* nomologically possible since, for all we have said so far, it might be that time travel is logically impossible, or metaphysically impossible. If so, then it must also be nomologically impossible. So in order to know whether time travel is nomologically possible we need to know whether it is logically possible, and, if it is, whether it is metaphysically possible.

8.2. Is Time Travel Logically Possible?

Let's start by considering whether time travel is logically possible. On the face of it you might think it's obvious that time travel is logically possible: after all, there doesn't seem to be anything contradictory in the idea of travelling backwards in time. Yet until Lewis' 1976 paper, it was often assumed that backwards time travel is logically impossible. The reason for this is that backwards time travel was thought to be something that could bring about contradictions, and since contradictions are impossible, so too, it was thought, time travel must be impossible. The grandfather paradox argument (see below) is one instance of an argument of this kind, designed to show that time travel is logically impossible.

The Grandfather Paradox Argument

[P1] Time travel is logically possible (assumption to be rejected).

[P2] If time travel is possible, then there is a possible time traveller who travels back in time and kills his grandfather before his father is conceived.

[P3] If the time traveller's grandfather is killed before his father is conceived then the time traveller's father is never born, and so neither is the time traveller.

[P4] So if the time traveller kills his grandfather before his father is conceived, then the time traveller does not exist.

[P5] If the time traveller kills his grandfather before his father is conceived, then the time traveller exists.

Therefore,

[P6] If time travel is possible, there is a possible time traveller who both exists and does not exist.

[P7] There is no possible time traveller who both exists and does not exist.

Therefore,

[P8] Time travel is not possible.

The argument appears compelling. It is surely true that if the time traveller travels back in time and succeeds in shooting his grandfather – call him *young grandfather* – before his father is conceived then the time traveller both exists and fails to exist. The time traveller exists because he shoots young grandfather and non-existent things cannot shoot anything. On the other hand, if the time traveller shoots young grandfather, then he himself will never be born, and hence he does not exist. So premises [3] through [7] look plausible. But [P2] also looks plausible. On the assumption that even a few time travellers travel back in time and try to kill their young grandfathers, it seems pretty likely that at least one of them will succeed. Hence we have a demonstration of the claim that time travel is impossible.

In response, Lewis argues that the grandfather paradox argument does not show that time travel is impossible. Rather, it shows that no one kills young grandfather. Lewis thus denies [P2] of the argument. It does not follow, he says, from the fact that time travel is possible, that it is possible for anyone to kill their young grandfather. For there is another option: the time traveller simply fails to kill their young grandfather whenever they try. What happens if a time traveller attempts to kill their young grandfather? Who knows? What we do know is that the attempts fail. Either the time traveller loses their nerve, or the gun jams, or the traveller suffers a stroke, or is prevented by a passing police officer, or the traveller accidentally kills the wrong person, or any of a host of other things happen; but what does not happen is a successful killing of young grandfather. But there is nothing contradictory about this: people try, and fail, to kill other people all the time. So the grandfather paradox argument merely shows that no time traveller ever brings about a contradiction; but *of course* that is true.

One might remain worried. After all, what motivates [P2] in the first place is the fact that it seems as though each of us *can* kill our own young grandfathers if we try. Indeed, it seems that each of us is as able to kill our own young grandfather just as easily as we might kill some random stranger at the mall. And if each of us can do something, we expect that many of us will succeed in doing that thing. If we collectively went out to malls and attempted to kill young people we would expect some of us to succeed. So too we expect that if we travel back in time to kill our young grandfathers, we would kill some of them. But according to Lewis anyone who attempts to kill their young grandfather has to fail. There are just some things that time travellers can't do, and killing their grandfathers before their fathers were conceived is one of those things. If that is right, however, then it would seem we both can and cannot kill our grandfathers. We can kill our young grandfathers in just the same way that we can kill anyone; we cannot kill our young grandfathers because of the peculiarities of time travel. But that's another paradox! So something has gone wrong.

Lewis' response to this worry is to distinguish two senses of 'can'. There is, he says, a sense in which each of us can kill young grandfather. Consider how we usually assess what it is that each of us can do. We consider our capacities, at a time, and this tells us whether we can do something or not. So, Sara can eat toast rather than cornflakes tomorrow for breakfast (assuming both are available) because she is fully capable of turning bread into toast and eating it. In this sense of 'can', each of us can kill young grandfather. Notice that when we are assessing what we can do, we do not include in the assessment future facts about what actually happens. Suppose we are assessing whether Sara can eat toast tomorrow morning for breakfast. Suppose also that, in fact, tomorrow she eats cornflakes. If we ask whether she can eat toast tomorrow *given that she eats cornflakes*, the answer would seem to be no. Given that she eats cornflakes, she can't eat toast. Her eating toast is not consistent with her eating cornflakes.

Likewise, Sara's killing young grandfather is not consistent with young grandfather living to a ripe old age and fathering her father, who fathers her. So, conditional on her grandfather not dying as a youth, she cannot kill him, just as conditional on her eating cornflakes tomorrow, she can't eat toast. Nevertheless, in the ordinary sense of 'can' she can kill young grandfather, just as she can eat toast tomorrow. It is just that in fact, in both cases, she doesn't: she does not eat toast, and she does not kill young grandfather. So relative to one set of facts, she can kill baby grandfather, and relative to another, she cannot. So there is no further paradox concerning the concept of 'can'. More importantly, the relevant set of facts is really the set according to which she can kill young grandfather, because that's the set of facts that we typically use to assess claims about what she can do. She can kill young grandfather, but she won't.

We have, then, an explanation for why it is that [P2] seems to be true. [P2] trades on the idea that because each of us can kill young grandfather, in the ordinary sense of 'can', then some of us would, in fact, kill baby grandfather. But therein lies the mistake, argues Lewis. Each of us can, but none of us will. Still, one might still be left with a queasy feeling. For one might think that what Lewis tells us about the usual sense of 'can' is just misleading. One might think that the usual sense of 'can' is one in which, if one can do something, then if one were to try sufficiently often, one would, or might, succeed in doing that thing. So, for instance, Sara can eat toast for breakfast. Notice that were she to try to eat toast, then she either will, or might, succeed. Equally, if she kept trying to eat toast, and kept failing, she would surely eventually conclude that she cannot eat toast. She would conclude that she is not free to eat toast. So a remaining worry has it that even if what Lewis has shown is that time travel is possible, there is still a puzzle about time travel and free will. For killing young grandfather is not like eating toast at all. No matter how many times any of us tries to kill young grandfather, we will fail. This raises the problem of freedom, to which we return in section 8.5.

8.2.1. The Second Time Around Fallacy

It is tempting to conclude that although no time traveller *will* succeed in killing young grandfather, a time traveller *might* succeed in killing someone to whom he is not related, even if that person did not die at that time, in that manner, the 'first time around'. Here is the thought. Suppose that the first time around, Frank grew up in the 1940s, lived to a ripe old age of eighty-nine, and died a peaceful death in 2016. Suppose, too, that Melanie, a time traveller, married Frank's son, Dave, and that Melanie very much regrets this decision. Divorce seems like a lot of hassle, so instead she decides to travel back in time and kill Frank before Dave is conceived. Imagine that Melanie succeeds in killing Frank and therefore succeeds in making it the case that Dave is never conceived. Then the first time around, before the time travel, Frank lived to a ripe old age, and the second time around, after Melanie travelled to the past, Frank dies young and childless. Since Melanie is not related to Frank, her killing of Frank does not seem to generate any paradox.

Nevertheless, many philosophers think that what we just described is impossible. Such philosophers think that changing the past in this fashion is impossible, and that the story we just told commits the *second time around fallacy*. It is a fallacy, on this view, because there is no second time around. There is no way things were, the first time, prior to the time traveller arriving in the past, and a different way things are, the second time, after the time traveller arrives in the past. Why so?

In order for something to change, it needs to be one way and then some other way. In ordinary cases of change things change when they are one way at one time, and some other way at some other time (this is the at-at conception of change discussed in Chapter 1). For example, suppose that on Monday at 2pm Annie's mass is 21.2kg (Annie is a dog). Suppose she wants to change her mass. She won't do this by trying to make her mass different *at Monday at 2pm*. Instead, she will try to make her mass less, or more, than 21.2kg at times later than Monday at 2pm. But notice that in order to change the past we need to change some past time from being one in which Frank is alive to being one in which Frank is dead. How do we do that? The obvious suggestion is that a time needs to go from being one way at one time to being another way at another time. But no time can itself undergo change. Change is what happens to things *in* time, by things being different *at* different times. Times themselves cannot change their properties, for there is no dimension along which they can change.

To be sure, things that *look* like changing the past are possible. It's possible to travel to a parallel universe that is like ours up until a time that is a past time in our universe, and to act at that time so that the future in that universe is different than the future in our universe. So it's possible to travel to a universe much like ours, kill a baby called 'Adolf' and, in so doing, make it the case that in that universe there is no Second World War. That, however, is not changing the past, it is just moving from a universe in which Hitler grew up and started the Second World War, to a universe in which he is killed as a child: our universe remains one in which Hitler was not killed as a baby. With this in mind, we can now construct a new argument against the possibility of time travel.

The Argument from Changing the Past
[P1] It is impossible for some past time, t, to change from being a time at which P to being a time at which not P.
[P2] Therefore there is no possible time traveller who changes some past time, t, from being a time at which P to being a time at which not P.
[P3] If travelling to the past is possible, then there is a possible time traveller – Fred – who travels to a past time, t, and causally interacts with some objects that exist at t.
[P4] If Fred travels to t, and causally interacts with some objects that exist at t, then Fred changes t from being a time at which P to a time at which not P.
Therefore,
[P5] If travelling to the past is possible, then there is a possible time traveller who travels to a past time and changes that time from being one at which P to a time at which not P.
Therefore,
[P6] Travelling to the past is not possible.

Here is a way of thinking about what is going on here. If Fred is a physical object travelling back in time in anything like a world like ours, then he will causally interact with objects at that past time. Even if he tries to be a mere bystander at that past time, he will breathe oxygen, create gravitational changes, refract light, and so on. But if Fred causally interacts with objects at the time to which he travels, then he will change the way things are at the time to which he travels. But it is impossible to change a time. So travelling to the past is not possible.

Does this argument show that time travel to the past is impossible?

No. That's because [P4] is false. Though it appears plausible, it mixes up causally influencing the past with changing the past. Let's see how.

8.2.2. *Influencing versus Changing the Past*

[P4] says that if Fred travels to some past time and causally interacts with objects at that time, then Fred changes that time from being one way to being some other way. That looks plausible. When Fred gets out of the time machine and treads on an ant, it seems as though he changes the time from being one in which the ant lived a long and productive life to one in which the ant died and was subsequently eaten by a passing bird.

Fortunately, we can accommodate the thought that Fred is *causally efficacious* in the past – that Fred can causally affect the time to which he travels by, for instance, treading on ants – without conceding that by doing so, Fred changes the past. Why doesn't Fred change the past when he treads on the ant? Because there is no first way that the time is, with an untrodden-on ant. There is only one way the time is: which is a time at which the ant is trodden on. It will help to make this clearer if we focus on a block universe model of temporal ontology.

Remember that according to the block universe theorist the entire four-dimensional block of events that did, does, and will occur exists, and that block does not change. Things in the block change, by being one way at one location in the block, and another way at a different location, but the block as a whole is static.

Imagine you are looking down at the whole four-dimensional block (you will need to imagine that you exist outside of time and space, see Figure 18). What do you see occurring at the time to which Fred travels? You see that Fred exits a time machine and treads on an ant. Fred's being there at that time makes a causal difference to that time. Had Fred not travelled back in time, the ant in question would not have been trodden on. So Fred being located at the time to which he travels is part of what makes that time the way that it is, namely, a way in which an ant is trodden on (and so forth). But Fred doesn't *change* that time from being a time with a live ant, to being a time with a dead ant. There is only ever one way that time is: a way with a dead ant.

We've used the block universe theory as a way to spell out how to understand influencing the past assuming there is no second time around. But we can do the same for other models of time. Consider presentism. If Fred is about to get into the time machine in the present, and if he does successfully travel back to the relevant time, then it is *now* true (before he

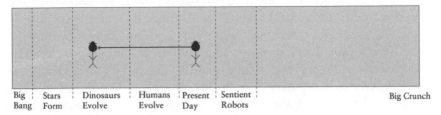

| Big | Stars | Dinosaurs | Humans | Present | Sentient | | Big Crunch |
| Bang | Form | Evolve | Evolve | Day | Robots | | |

Figure 18 Time Travel in a Block Universe: This figure depicts a block universe with a time traveller, mapping their trajectory, across time.

gets into the machine) that Fred *was* located at that earlier time, and that at that earlier time he *did* step on the ant. To be sure, the events at that earlier time no longer exist. But they *did* exist, and when Fred was there, he did step on the ant. Nothing he does now, by getting into the machine, changes what *did* happen at that earlier time. We leave it as an exercise for the reader to explain how other models of time (such as the growing block and moving spotlight) might understand influencing the past without committing to there being a second time around.

It's important to notice that just because there is no second time around, it doesn't follow that the time traveller does not influence the time to which she travels. Had things gone differently – had the time traveller decided not to travel, for instance – then what happened at the earlier time would have been different: the time traveller would not have been present. So even if it is impossible to *change* the past, we can make sense of the causal interaction of time travellers with the past.

8.2.3. Hypertime and Changing the Past

In the previous section we noted that most philosophers think it is impossible to change the past. Unless one posits a second temporal dimension, there is no dimension along which a time can change from being one way to being some other way. If, however, one does posit a second temporal dimension – such as we met in Chapter 2, i.e. hypertime – then perhaps one can make sense of the idea that a time is one way at one hypertime, and another way at some other hypertime. Let us say a little more about how that works, and whether, if it does, it really counts as changing the past. Let's start by considering a two-dimensional model of changing the past, and let's assume an eternalist B-theoretic model of temporal ontology. Now, let us say that a time, t, can have one set of properties at one hypertime, and some other set of properties at a different hypertime. To understand how this works we need to introduce not only times (designated by t) but also hypertimes (designed by ht). In a world without time travel, time and hypertime are in sync. A given time t_7 will match up to the corresponding hypertime ht_7. Once there is time travel, however, times and hypertimes are no longer in sync. Relative to hypertime, a time traveller moves forwards, even if she moves backwards in ordinary time. So, for instance, if Tim steps into a time machine and travels backwards in ordinary time from t_4 to t_1, he nevertheless travels forwards in hypertime. So, suppose that Tim aims to change the past (t_1) to which he travels and then return to the present time (t_4) (see Figure 19).

At t_1, ht_1, we have the 'original' version of t_1, at which there was, say, a war. At t_4, ht_4, Tim steps into the time machine and travels back to t_1. But the t_1 to which he travels is t_1, $ht_5 - t_1$ as it is, at a different hyper-temporal location. Tim influences t_1 (at ht_5) preventing the war, and then travels back to t_4 at ht_8. The t_4 to which he returns is different than the t_4 from which he travelled (since it is the result of there having been no war). We can make sense both of the change in t_1, and the change in t_4, by noticing that the changed t_1 and t_4 occur at different hypertimes.

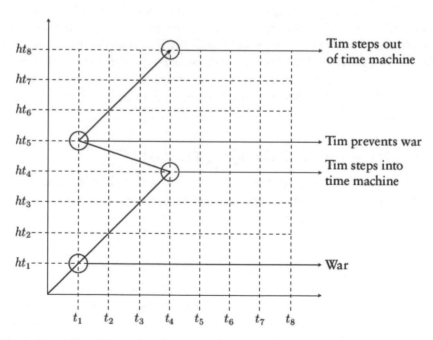

Figure 19 A Two-Dimensional Time Travel Story: The horizontal axis represents time, the vertical axis represents hypertime. Each point in the 2D space is a time, hypertime pair.

Once we introduce time travel into a model with two-dimensional time we need a complicated two-dimensional diagram, such as Figure 19, to keep track of the various ways times are, at hypertimes. Importantly, not only does the diagram show how to make sense of influencing t_1, but it seems to show why this really is a model of changing the past. After all, when Tim returns to t_4 (at ht_8) everything is different: there was no war, in the past, according to t_4, ht_8. No one at t_4, ht_8 remembers a war; there are no records of there having been a war, and so on.

It is controversial whether this really does amount to *changing* the past, or, as some philosophers have argued, only to *avoiding* the past. Here is why. What would it take to change the past? One might think that to change the past means to erase the ways things were, in the past, and replace them with some other way things were. Suppose, for instance, one remembers the terrible war that occurred at t_1. The aim of changing the past, one might assume, is to erase that event and replace it with some other. That, however, is clearly not what happens in models such as the ones we are discussing. For at t_1, ht_1 the war occurs. That never changes.

In models such as these, times have hyper-temporal parts: times themselves are 'extended' along the hyper-temporal dimension. That means that if a hyper-temporal part of a time is one at which a war occurs, then it is always the case that that hyper-temporal part is one at which a war occurs. To put it another way, the two-dimensional diagram does not change: it is static, just like a block universe is static. If an event occurs at a time/hypertime pair, then it occurs at that pair, and nothing can be done to change that. All that can be done is to create a new 'version' of the time, at a different hypertime (t_1 at ht_5) that is different from that time at some other hyper-temporal location (t_1 at ht_1). But that is not changing t_1 in any sense we care about, since the original t_1 at ht_1 is still there, as it always was.

We can make worries such as these a bit more concrete by thinking about what we might mean by 'the past' in the context of such models. Consider the two most obvious options. The first option says that what is past, relative to some time t, is the set of time/ hypertime pairs that have as their first member any time that is earlier than, or simultaneous with, t. So relative to t_4, in our diagram, the past will be the shaded region in Figure 20. We can easily see why, if this is what we mean by 'the past', it will be impossible to change the past. For if an event occurs at some past time/hypertime pair, then that event occurs at that pair: nothing any time traveller does will alter that.

The second option would be to suppose that what is past relative to a time is, in part, determined by which hyper-temporal part of that time we

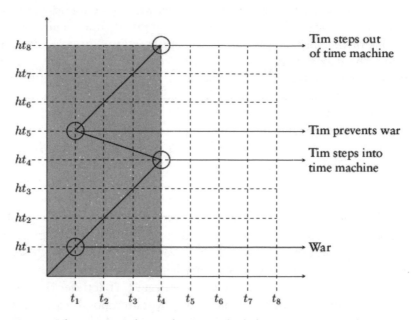

Figure 20 The Past as a Sheet: The entire shaded region corresponds to the two-dimensional past.

are attending to. Consider t_4 at ht_4. We could say that the past with respect to that time/hypertime pair is the set of time/hypertime pairs that include only: t_0, ht_0; t_1, h_1; t_2, ht_2; t_3, ht_3. By contrast, we could say that the past relative to t_4 ht_8 includes t_0, ht_4; t_1, ht_5; t_2, ht_6; t_3, ht_8. Then the two different pasts of t_4, relative to each hyper-temporal moment, can be depicted as in Figure 21.

On this way of defining the past, then, the past relative to some time/ hypertime pair, t, ht, is the set of time/hypertime pairs that includes all pairs in which the time is earlier than t and the hypertime is earlier than ht. But if this is how we think of the past then Tim fails to change the past. Relative to t_4, ht_4, it was and always will be true that the past is one in which the war occurred. Tim does not change that. And relative to t_4, ht_8, the past is one in which the war does not occur. But that is, and always will be, true. Tim does not change anything. Nothing goes from having been one way to being some other way. Instead, we simply have two sets of pasts, with Tim travelling between them. He avoids Past 1 by travelling to Past 2, but he *changes* nothing.

Whether two-dimensional accounts really model changing the past depends, then, on what one thinks it would be to change the past.

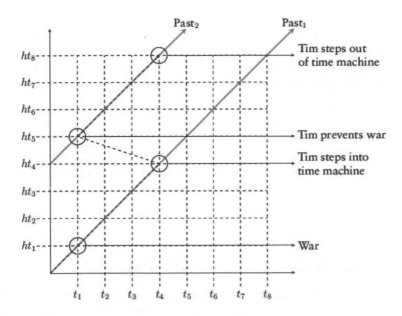

Figure 21 The Past as a Trajectory: Each line represents a different two-dimensional past.

Defenders of these models point out that in a block universe we are happy to think that objects change by being one way at one time and some other way at some other time: we do not require that the way they were, at the earlier time, has vanished from existence or somehow been 'overwritten' by the way things are later. Thus one might argue that when times bifurcate by having hyper-temporal parts, as they do when there is time travel, the objects that exist at those times are the very same objects. If Frank is walking down the street at t_1, ht_1, then it is the very same Frank walking down the street at t_1, ht_4. If Tim kills Frank at t_1, ht_4 then Tim changes the past since he brings it about that Frank – the very same Frank – dies at t_1, ht_4. It cannot be required that Frank as he is at t_1, ht_1, is also killed, for that would be contradictory. We won't adjudicate on this issue any further here.

It's also worth noting that even if the past really does change on such models, they require that we posit hypertimes. But that is an ontological cost; moreover, there is no evidence for the (actual) existence of hyper-times. So one might complain that positing hypertimes (for any purpose) is really just ad hoc. We have no reason to posit such things other than that we want to accommodate changing the past: but that's not a reason to posit a whole additional temporal dimension! This is worth bearing in mind since, as we will see, more recently somewhat different models of changing the past within a dynamic temporal ontology have been offered. What both sets of models share, however, is an appeal to hypertimes. If one thinks that positing hypertimes is unmotivated, then regardless of whether one thinks that these models *do* model changing the past, one will think that, at best, changing the past is logically possible (but not actual), and, at worst, one will think that there are no worlds with hypertimes, and hence no world where in fact the past is changed. We will return to this issue shortly. For now, we will end our discussion of the logical possibility of time travel here. In what follows we move on to consider whether time travel is metaphysically possible.

8.3. Is Time Travel Metaphysically Possible?

So far we have asked whether there is anything about the nature of time travel itself that is contradictory, and in virtue of which time travel is logically impossible. We failed to find any such contradiction. That, however, does not show that time travel is metaphysically possible. For it could still be that time travel is inconsistent with our best metaphysical theories of the world. Perhaps, for instance, presentism is true, and true of metaphysical necessity. Then if time travel is inconsistent with presentism, time travel is metaphysically impossible. It is this, and possibilities like it, that we shall now consider.

8.3.1. Presentism and Time Travel

Recall that on an eternalist temporal model, past, present and future objects exist, as do past, present and future times. According to eternalism, just as spatial locations other than the one that you or I are located at exist, so too temporal locations other than those at which you or I are located exist. On the face of it, then, it seems plausible to suppose that, in principle, just as you and I can travel to other spatial locations, so too you and I can travel to other temporal locations.

Suppose, however, that presentism is true. Presentism is the view that only present concrete entities exist. So it is easy to see why one might be sceptical that presentism is compatible with time travel. After all, it seems that there are no past concrete locations to which to travel. This objection is known as the no destination argument.

The No Destination Argument
[P1] A time traveller can only travel to a location, L, if L exists.
[P2] If presentism is true, past locations do not exist.
Therefore,
[C] If presentism is true, no time traveller can travel to any past location.

If the no destination argument is sound, and if presentism is true and metaphysically necessary, then time travel is metaphysically impossible.

The argument, however, faces a problem. Suppose we replace 'past' with 'future' throughout the argument. Then the argument shows that we cannot travel into the future either, since according to presentism future locations do not exist. But presentists think we are travelling into the future as we speak. So either this argument is really an argument against presentism itself – since it shows that time cannot flow the way the presentist says it does – or the argument is unsound. In either case, it doesn't show that time travel is incompatible with presentism. What should the presentist say about travel into the future? How do we manage to travel from this moment to the next moment, a moment that is, at present, a future and hence non-existent moment? The presentist will say that we can do this because by the time we get to that future moment, it is no longer future but present, and hence exists. Consider a spatial analogy. Suppose some spatial location – call it non-existent land – does not presently exist, but someone is selling tickets to travel to it. Suppose, though, that the tickets specify that the journey will take ten years. At the end of ten years, we are assured, non-existent land will exist because the spacetime that will constitute non-existent land is being created as we speak. The presentist should insist that travelling to a future time is like travelling to non-existent land. The future time does not exist *now*, but it *will* exist, at the point where we reach it. But if the presentist can say that about a future time, then she can also say it about a past time. To be sure, no past time *now* exists. But all that matters is that said time *used to* exist; that the past time was there, when the time traveller arrived.

So it seems that the presentist has a good response to these kinds of objections. There is no reason to think that presentism is incompatible with time travel, and thus no reason to think that even if presentism is metaphysically necessary, time travel is metaphysically impossible.

8.3.2. Moving the Present to Change the Past

We have already considered time travel in a presentist world. It is worth noting that while it is generally agreed that we can tell a coherent time travel story that is consistent with presentism, not everyone is so sure that this ought really to count as time *travel*. After all, what really happens? The traveller is located in the now – the only moment at which objects can exist. By getting into a time machine now, the time traveller makes it the case that some past-tensed statement is true. She makes it true, for instance, that she *existed* at some past time. The objects that existed at that time, however, including the time traveller herself, *no longer exist*. So if the time traveller gets into the time machine and flips the switch, she brings it about that she ceases to exist. To be sure, she also brings it about that she *did exist*, at some earlier time, but in so doing she also brings it about that she *does not exist*. That is because the time to which the traveller travels is objectively past, and hence the objects that existed at that time no longer exist. Time marches ever onwards regardless of the actions of the time traveller. If t_{20} is the objectively present moment, then the next objectively present moment will be t_{21}, regardless of the fact that the time traveller travels to t_5.

Recent models of time travel have sought to avoid this outcome by holding that the time traveller effectively *takes the present moment with her* when she travels. So, suppose we are considering Frederika, who travels from t_5 to t_3. In order to understand how this works we need to invoke hypertimes. As before, hypertime always unfolds chronologically. And, again, in the absence of time travel, hypertime and time are always in sync. Time travel, however, consists in moving the objectively present moment; in effect, this is to re-wind time. We can suppose that Frederika was not, originally, i.e. the first time around, at t_3. But by travelling there she re-winds time so that t_3 is the objectively present moment, and she exists at t_3. In hypertime then, the order of times is as follows: t_1, t_2, t_3, t_4, t_5, t_3, t_4, t_5. Hypertime always unfolds chronologically, ht_1, ht_2, ht_3, ht_4, ht_5, ht_6, ht_7, so the order of times in hypertime is as follows: ht_1, t_1; ht_2, t_2; ht_3, t_3; ht_4, t_4; ht_5, t_5; ht_6, t_3; ht_7, t_4; ht_8, t_5.

Since there are multiple ways of modelling temporal dynamism there are multiple versions of this model of changing the past. We shall focus on just two: the presentist version and the growing block version. On the presentist version of the model only the present moment exists. By travelling to past times, the time traveller takes the present moment with her: no other times exist, other than the time to which she travels. So, at ht_7, only t_3 exists. On the growing block version of the model, by contrast, when the time traveller travels back in time she effectively erases all of the block from the location to which she travels, up until the time at which she departed. So if the traveller leaves in 2016, and arrives in 1970, she deletes the block between 1970 and 2016. Then a new block regrows from the new 1970. But the regrown block will be different from the original block, since the time traveller will causally affect the way things progress. Hence, she will have changed the past.

What both models share is the idea that because the traveller moves the objective present around, she can have a 'do-over' of time: there really is a second time around that is different from the first time around. There is no contradiction involved since the way things were, the first time around, has been erased from existence by the movement of the objective present. What this means is that different facts about the past will be true at different hypertimes. If Frederika travels back in time and goes to a wedding that she avoided the first time around, then there will be a hypertime ht_3, at which, in the past, Frederika did not attend the wedding, and a hypertime ht_6, at which, in the present, she does attend the wedding, and a hypertime ht_7, at which, in the past, she did attend the wedding.

Does this model accommodate changing the past? It's worth noticing that there are two ways we can think about hypertime in this model. First, we could think that hypertime is eternalist: all hyper-temporal events that did, do, or will exist, do exist. If we think of hypertime this way, then it is not clear that this model is really any better than the standard two-dimensional model we previously considered. Suppose that the growing block model is right, and we consider Frederika and her wedding-attending behaviour. Suppose that at t_3, the first time around, Frederika fails to attend the wedding, and that is why, at t_5, she travels back to t_3 in order to attend the wedding. So the first time around at t_3 Frederika is absent from the wedding, and the second time around at t_3 Frederika is present at the wedding. That means that at some hypertime (namely ht_3) Frederika is absent at the wedding at t_3, and at some other hypertime, namely ht_6, Frederika is present at the wedding at t_3. If the hyper-temporal dimension is eternalist – all hypertimes are equally real – then we can represent this as in Figure 22. For our purposes what matters is that the *absence* of Frederika at the wedding at t_3 still exists at hypertime ht_3. To be sure, it no longer exists in time, at ht_6: that slice of reality has been deleted. But it is still true, at ht_6, that at an earlier moment *in hypertime*, t_3 (a time without the wedding attendance) exists. The slice has been erased from time, but not hypertime. The same is true if presentism is true of ordinary time, but eternalism is true of hypertime. Then although at ht_3 the only time that exists, t_1, is one in which there is wedding attendance, there is some earlier hypertime, ht_1, in which the only time that exists, t_1, is a time at which there is no wedding attendance. In general, if we pair any dynamic theory of ordinary time with a non-dynamic, eternalist model of hypertime, we get a model that shares many of the features of the two-dimensional models we have already considered. If the two-dimensional model we previously considered cannot accommodate genuinely changing the past, then it is hard to see how these models do any better.

The only way to avoid this consequence would be to hold that hypertime is presentist. Then the only hypertime that exists, let us suppose, is ht_6, and hence there is no record at all of Frederika's failure to attend the wedding at ht_3, t_3. For the sum totality of reality is represented by just the sub-portion of the diagram in Figure 22, which is the block that exists as of ht_6, since no other hypertimes exist. Of course, it will always be true that, at some earlier hypertime, it was true that, in the past, the wedding was not attended. *That* can never be changed: the past-time hyper-temporal facts are there to stay. But there is no record of these in ontology. They are, effectively, erased. If anything is to count as changing the past this will be it. We leave it to the reader to consider whether the resulting picture of time and of reality is any good.

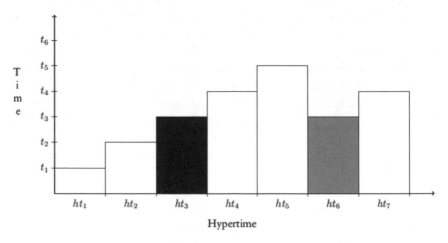

Figure 22 Dynamic Time, Hypertime and Time Travel: Hypertime is eternalist, and ordinary time is modelled by the growing block. At ht_3, t_3 Frederika fails to attend the wedding. At ht_6, t_3 Frederika attends the wedding. The block that exists at ht_3 and ht_6 are different. At ht_7, t_4 it is true that Frederika did attend the wedding at t_3.

8.4. Is Time Travel Improbable?

We turn now to consider whether time travel is *improbable*, despite being logically, metaphysically and nomologically possible. Why might we think time travel is improbable? Well, if time travel is probable, then lots of people are likely to travel in time. But if lots of people are travelling in time, we should expect quite a few of them to be trying to change the past in various ways, including, for instance, trying to kill their young grandfathers. All these attempts to change the past, and to kill young grandfather, will fail. So if time travellers are repeatedly engaging in these kinds of behaviours, we should see a whole range of failed attempts to do things like kill grandfathers. Indeed, for any past event, E, that a time traveller repeatedly attempts to change, there will be a series of failed repeated attempts to change E. So if enough time travellers are engaged in enough attempts to change the past, we expect to see long strings of coincidences in which individuals repeatedly fail to achieve their intended goals. We will, for instance, see a great many gun jammings, or sudden changes of heart, or cases of mistaken identity as each time traveller's murderous attempts are foiled. We do not, however, see these long strings of coincidences. Since there would be such strings of coincidences if time travel were probable, and since we don't see these coincidences, it follows that time travel is improbable.

> **The Argument from Improbability**
> [P1] If time travel is probable, then for some past events, E1 … En, there will be multiple attempts to change each of E1 … En.
> [P2] Each attempt to change each of E1 … En will fail because changing the past is impossible.
> [P3] As each attempt to change each of E1 … En fails, there will be a long string of coincidences.
> Therefore,
> [P4] If time travel is probable, we will see a long series of such coincidences.
> [P5] We do not see a long series of such coincidences.
> Therefore,
> [C] Time travel is not probable.

In response to the argument outlined above, one might begin by disputing the claim that if time travel is probable, lots of time travellers will try to change the past. After all, perhaps once time travel technology is developed, potential time travellers will be much more metaphysically sophisticated than we are, and they will all just agree that they will not succeed in changing the past, so they won't even try. If we think it is not likely that time travellers will try to change the past, then the above argument clearly fails to show that time travel is improbable. But suppose we think that it is likely that time travellers will attempt to change the past. Then should we conclude that time travel is improbable?

Probably not. Certainly, we have not, as yet, seen the long strings of coincidences that the argument appeals to: cases of mysterious failures to murder youthful grandpas. So if we think it likely that time travellers *will* try to change the past, there is something we can conclude: namely that time travellers have not travelled to anywhere near here (temporally speaking). Time travellers have not travelled to our local past, or to this time. For if we think it likely that time travellers will try to change the past, then, given that we do not see long strings of coincidences, we know that time travellers are not trying to change the past around here. But that does not establish that time travel is improbable. After all, there are very many locations to which time travellers can travel besides this one. So maybe they just haven't bothered to visit us yet. Or maybe they can't, because of some other constraint that they are under (perhaps it requires too much energy to get here).

The argument from improbability asks us to infer that because there are no strings of coincidences in our local past, there are no strings of coincidences anywhere, and therefore that it is improbable that there are time travellers anywhere. That inference may well be false. Consider the following analogy. Imagine Parramatta Road (a main road in Sydney) in the 1920s. Were someone to roll a whole bunch of tomatoes down Parramatta Road in 1920, it would be unlikely, and coincidental, if every tomato was hit, and squashed, by a car. That is because there was very little traffic in 1920, so most tomatoes should escape unharmed. Following the same line of reasoning used in the argument from improbability, though, we should infer from this that if we roll a bunch of tomatoes down Parramatta Road today, it will be coincidental if every one of them is squashed. That, however, is clearly a mistake, since there is a lot of traffic on Parramatta Road today. Indeed, it would be coincidental if any tomatoes, rolled down the road, failed to be squashed.

The point, then, is this. Tomato squashings in 1920 were improbable. But what is improbable depends on local conditions: in this case the absence of cars on the road. What is true for tomato squashings is true for so-called strings of coincidences. We cannot infer that there will not, in the future, be strings of coincidences, just because there were no such strings of coincidences in the past. After all, if future time travellers bent on changing the past begin to travel back in time to, say, ten years in our future, then in ten years we will begin to see strings of coincidences: except, of course, they will not seem coincidental any more. If lots of time travellers try to kill their young grandfathers, then in ten years' time we will expect to see a lot of failed attempts on the lives of grandfathers. Failed attempts such as these will become probable. So just as we cannot infer from the fact that tomato squashings were rare in 1920 to the fact that they are rare now, and then to the fact that there will be little traffic on the roads now, so too we cannot infer from the fact that there are few failed grandfather-killing attempts in the local past, to the fact that there will be few failed grandfather killings in the future, to the fact that there will be no time travellers located in the future trying to kill their young grandfathers.

So we have found no reason to think that time travel is improbable. At best we have reason to think that there are no time travellers in our local past, hell bent on changing the past in various ways.

8.5. Time Travel, Free Will and Deliberation

So far we've considered whether time travel is logically, metaphysically or nomologically impossible, and whether it is improbable. But time travel might be possible, and indeed probable, and still raise difficult metaphysical issues. In particular, one might think that if it is indeed impossible to change the past, then various interesting issues are raised regarding how time travellers ought to deliberate about what to do in the past.

8.5.1. Are Time Travellers Free?

According to the analysis of 'can' discussed in section 8.2, a person can do something if doing that thing is compatible with their capacities. But perhaps that is not the right account of 'can'. A better account might be this: a person can do something, and hence is free to do that thing, only if, were they to try, they would, or might, succeed. Sara has, in fact, never tried to fly. What makes it true that she cannot fly, and that she is not free to fly, is that were she to try, it is not the case that she would, or might, succeed. How do we evaluate whether or not a person would, or might, succeed in doing something that they have not, in fact, ever attempted to do? We have already met the answer to this question in chapter 6, when we appealed to counterfactual conditionals. Recall that the claim that were Sara to do X, Y would happen (even though Sara doesn't, in fact, do X), is a *counterfactual conditional*. We noted, in chapter 6, that a common way to evaluate counterfactuals involves imagining what happens at worlds where the antecedent of the counterfactual is true. So, for instance, consider the counterfactual: if Yufei were to drive at 200 miles per hour, she would crash. This is true if, amongst the worlds in which Yufei does drive at 200 miles per hour, there is no world in which she fails to crash, which is *closer* to the actual world than any world in which Yufei drives at that speed and does crash.

So let's return to our worry about freedom. Suppose that a person is able to do something if, were they to attempt to do it, they would, or might, succeed. Then we could say that a person might succeed in doing something only if there is some nomologically possible world in which they succeed in doing that thing. So, consider whether Sara can eat toast for breakfast. She can, on this view, since even if in fact she has never attempted to do so, there are lots of nomologically possible worlds in which she attempts to eat toast and succeeds. By contrast, consider the question of whether Sara can fly. Clearly, Sara can't, because there is no nomologically possible world in which she tries to fly and succeeds. It might be thought that the same broad story applies to killing young grandfather. For there is no nomologically possible world in which any of us attempts to kill young grandfather and succeeds. (We will succeed in very distant worlds in which babies can be resurrected, but that is irrelevant for the purposes of determining whether or not any of us *can* kill young grandfather.)

If the broad approach to 'can' discussed above is right, then it shows that although time travel is possible, the freedom of time travellers to act in the past is constrained in ways that it is not constrained in the present. After all, there are nomologically possible worlds in which, for each of us, if we attempt to kill our *ageing* grandfathers then we succeed. So it is now true, for each of us, that we can kill our ageing grandfathers. Thus we are now free to do things that we are not free to do in the past.

But is it correct to say that we *can* kill ageing grandfather, but *cannot* kill young grandfather? Let us think a bit more closely about how we go about assessing counterfactuals. Think about the case of breakfast. We want to know whether there is a nomologically possible world in which, when Sara attempts to eat toast, she succeeds. Suppose, though, that we asked the following question: is there any nomologically possible world in which, given that Sara in fact eats cornflakes, she eats toast? No. In any world (and thus in the nomologically possible ones), if Sara eats cornflakes at a particular time, she doesn't eat toast at that time.

To make this point clearer, consider what we might call the class of permanent bachelors. This is a class of people who, by definition, remain bachelors their entire lives. Suppose Jeff is a member of that class. We can ask whether Jeff *could* have married. The answer is surely yes: there are nomologically possible worlds in which Jeff decides to marry, and succeeds. So he can marry. Indeed, for each person in the class of permanent bachelors, surely each of them can marry: there is no powerful anti-marriage demon preventing each of them from marrying. But of course, if Jeff marries, then he is no longer in the class of permanent bachelors. What it is to be a member of that class is to fail to marry. But suppose we ask whether, *given that Jeff is a permanent bachelor*, he can marry. Well there is no nomologically possible world in which Jeff is a permanent bachelor and he succeeds in marrying. Indeed there is no possible world in which Jeff is a permanent bachelor and he marries. Yet we would hardly conclude, from this, that Jeff is not free to marry. We would conclude that when we assess whether Jeff can marry, we ought to determine whether or not there is a nomologically possible world in which he marries, not whether there is a nomologically possible world in which he is both a permanent bachelor *and* marries.

The mistake we make when assessing whether Jeff can marry, if we hold fixed that he is a permanent bachelor, is the same mistake we make when we assess counterfactuals about killing young grandfather. Or so goes the thought. Here is why. What it is to be a permanent bachelor is to fail to marry. So when we ask about Jeff's capacity to marry, we shouldn't hold fixed that he is a permanent bachelor, for then we hold fixed that Jeff fails. Likewise, when we ask whether time traveller Fred can kill young grandfather, we are illicitly holding fixed the relations between Fred and the young man he attempts to kill, and therefore holding fixed that he fails. Suppose we simply ask: can Fred kill that young man? Then the answer is yes. For there is a nomologically possible world in which Fred kills that young man: that will be a world in which a man just like that one exists, but is not Fred's grandfather. So although there is no nomologically possible world in which Fred kills young grandfather, we should not conclude from this that Fred lacks freedom because he can't kill young grandfather. For we get the same result if instead we ask whether Jeff, the permanent bachelor, can marry. In each case, though, when we assess the correct counterfactual, we find that Jeff can marry, and that Fred can kill young grandfather.

8.5.2. Deliberating About the Past

We have already seen that philosophers distinguish changing the past from causally affecting the past, and that most hold that the former is impossible, and the latter possible. Suppose this is right. If time travellers can causally affect, but not change, the past, it seems reasonable to ask: what sorts of things ought time travellers try to do? Given that Tim's grandfather is alive in 2017, not only do we know that Tim will not kill his grandfather in 1980, but Tim is also in a position to know that he will not kill his grandfather in 1980. Here's where things get interesting. Plausibly, someone can only deliberate about whether to do X, if that individual does not know *whether or not X*. For instance, one cannot deliberate about whether to turn blue when one knows one won't turn blue. So suppose one knows exactly what happened on some day in the past, including knowing that one's future self time travels to that day, puts on a duck suit, and marches in the 'free the ducks' parade. Then one cannot deliberate about whether one will travel back in time to that day and put on the duck suit, any more than one can deliberate about whether to turn blue. That's because one knows what one will do, and knowing what one will do precludes one from deliberating about what to do.

It is, however, rare that we know everything about some past time. So that leaves plenty of room for potential time travellers to deliberate about what to do in the past. Some things, no doubt, will be ruled out as deliberation worthy. In so far as Sara is certain that Hitler grew to adulthood, she cannot deliberate about whether to travel back in time to kill him in his youth, since she knows that she does not succeed in doing any such thing.

Of course, we have greater and lesser amounts of evidence about the past. Sometimes evidence is weak. In cases where evidence is weak, or where the stakes are high, a traveller might rationally deliberate about acting in the past so as to bring about some desirable consequence. For instance, suppose a traveller has weak evidence that 10,000 people died of famine at some past time. Then she might decide to travel back to give them crop advice, reasoning that the weak evidence she has that they died of famine may be misleading, and that by travelling back in time and offering advice she might save 10,000 lives. After all, perhaps it is now true that 10,000 people did *not* die of famine, though they almost did, and perhaps that is true *because* the traveller narrowly managed to prevent the famine.

Indeed, some have wondered whether it is reasonable for time travellers to manufacture evidence about the past in precisely these sorts of cases.

Suppose that rather than there being weak evidence that there was a famine, there is really a fair amount of historical evidence. Then either the famine occurred, in which case there is no point in the traveller travelling back in time to try and prevent it, or the famine did not occur, perhaps because the traveller prevented it, and the evidence in the historical records is misleading. In so far as the traveller is motivated to try to make it the case that the 10,000 people did not die from famine, it seems that she has reason to try and fabricate historical evidence of a famine.

To see why, suppose the traveller is considering whether to travel back in time to try and prevent the famine. She has some reason not to do this, given her evidence that the famine occurred. But suppose she travels back to times *after* when the famine would have occurred, if it did, and instead tells a range of famous historians a story about a terrible famine that killed 10,000 people. If the traveller does this, then she has an excellent explanation for why there is evidence of a famine: namely her time travelling self is responsible for the evidence. But then the traveller has reason to think that the evidence of the famine having occurred is misleading. Since she will then not be sure that the famine did happen, it will be rational for her to decide to travel back in time to attempt to prevent the famine from happening.

So it seems that the traveller ought to first travel back in time and plant evidence that there was a famine, then travel further back and make sure the famine does not occur. But, you might think, it cannot be reasonable to manipulate evidence in this way: after all, either the famine happened or it didn't; either the traveller prevented it, or she failed to, and no amount of manipulating historical records will change that.

Interestingly, though, what is true of the past is also true of the future in at least the following respect: in so far as one came to have knowledge about the future, and in particular knowledge about what one will do in the future, then one would be unable to deliberate about what to do in the future. Of course, it is very much more difficult, and less likely, that one will have knowledge of what will happen in the future, or what one will do in the future. That's because while there are records of the past, there are not (typically) records of the future. But it's not in principle impossible to know what will happen in the future. One can imagine there being an oracle that is 100 per cent accurate in predicting the future, and who one day tells you what you will do tomorrow. Or one can imagine that one meets one's time travelling self from the future, who explains exactly what you will do tomorrow. One can even imagine computers becoming so good that they can predict exactly what will happen at some future time, by looking at the laws of nature and the current state of the world. In all of these cases one's source of future knowledge is different from one's usual source of past knowledge. So there are some genuine asymmetries between past and future knowledge. Our point is just that in so far as one did have future knowledge, one would be in the same position regarding deliberation about the future as one is in deliberating about the past. That's because one would find oneself unable to deliberate about what one would do in the future under those circumstances.

But now suppose that what you come to know about the future is not what you will do, but rather evidence about what you will do. For instance, perhaps you come to have evidence that tomorrow you will rob a bank, because the oracle (or your future self) gives you evidence that tomorrow you are wanted for bank robbery. So you come to know that tomorrow you will be wanted for bank robbery. What you do not know is *why* you are wanted for bank robbery. One possibility is that you are wanted for bank robbery because you robbed a bank! Now notice that the more certain you are that you *will* rob the bank tomorrow, the less it will make sense to deliberate about *whether or not* to rob the bank. Moreover, just as it makes sense for the time traveller to create false evidence of a famine, it also seems to make sense for you to try and create false evidence of you committing a bank robbery (perhaps by travelling into the future and planting such evidence). For once you know that such false evidence exists, you have good reason to think that the evidence you saw is false, and therefore good reason to think that the reason there is such evidence is that you manufactured it, rather than it being there because you robbed the bank. So it then becomes reasonable to deliberate about whether or not to rob the bank, since you have undermined any evidence that you will, in fact, rob the bank. In this regard, the past and future are symmetrical.

It doesn't really matter whether one is travelling into the past or the future; what matters is how much the traveller knows about the time to which she travels. If the traveller knows a lot about what she did, or will do, then her capacity to deliberate about what to do is severely undermined. But if the traveller does not know a lot about what she did, or will do, then her capacity to deliberate remains intact. And if a traveller thinks she knows what she did, or will do, and would prefer that this is not what did, or will, happen, then perhaps it is rational for her to try to create misleading evidence *as of* that thing happening, in order to undermine her own apparent knowledge of that thing happening. That's pretty weird. That's not the only respect in which time travel gets pretty weird.

8.6. Time Travel and Explanation

The possibility of time travel seems to raise certain puzzles for explanation. Suppose an art critic travels back in time with a copy of an artist's masterpiece, gives the artist the copy, and the artist copies it. The copy of the copy turns out to be the original masterpiece. The artist comes to paint the masterpiece in virtue of seeing a future copy of that masterpiece and then copying it. Call the masterpiece 'original' and the copy-from-the-future, 'copy'. The scenario is puzzling because the artwork seems to come from nowhere. The artist never comes up with the idea for the painting: he simply copies it. Moreover, the copy only exists because the original exists – since it is a copy of the original. But so, too, the original only exists because the copy does, because the original just is a copy of the copy. What is perplexing about causal loops such as this is that something seems to be created from nothing, or *ex nihilo*.

There is at least one respect in which this case is not puzzling. Namely, we can provide a full causal history of original and copy. Why does the copy have the properties that it does? Because the original has the properties it does, and the copy is a copy of the original. Had the original been different, the copy would have been different. Since the original is a copy of the copy, it is also true that had the copy been different, the original would have been different. If we want to know why each is an oil painting we have an answer: they are both oil paintings because the copy is a copy of the original, which is an oil painting, and the original is a copy of the copy, which is an oil painting. That still leaves some puzzles. Suppose we want to know why *both* are oil paintings rather than, say, watercolour paintings. Certainly we can say that they are both oils because given that the original is oil, then the copy will be oil, and given that the copy is oil, then the original will be oil. We have a story about why they are relevantly similar in these respects. But it is less clear that we have a story about why they are *both* oils rather than watercolours. What is puzzling is not any particular link in the causal chain, but the entire causal loop.

In the scenario in question we seem to have what is known as a *closed causal loop*. A causal loop is *closed* if every event on the loop is both a cause and an effect of some other event in the loop, and no event on the loop is either a cause or an effect of any event that is not in the loop. Closed causal loops have no causal history outside of the loop: they are completely causally disconnected from all events that are not in the loop. We can contrast closed causal loops with open causal loops. These are loops in which at least some of the events on the loop are caused by events outside the loop (as well as causing events on the loop). So the loop itself has causal connections to things outside it. Open loops don't seem to pose the sort of explanatory gaps that closed loops do. That's because in an open loop we can explain some event in the loop in terms of events outside the loop, and then we can explain the rest of the events in the loop by appealing to the event that is causally connected outside the loop. By contrast, in a closed loop, while every event in the loop is explained by some other event in the loop, since none of the events in the loop is causally connected to any event outside the loop, nothing seems to explain the existence of the loop itself. While this is mysterious, however, it doesn't show that time travel is impossible: it just shows that some time travel scenarios are weird.

8.7. Summary

In this chapter we have discussed the paradoxes of time travel. The central points to take away from this chapter are these:

(1) Time travel involves a disparity between personal time and external time (either in terms of duration, or in terms of temporal structure).
(2) Time travel is to be differentiated from persistence through time.
(3) Time travel can involve travel into the past (backwards time travel) or travel into the future (forwards time travel).
(4) The grandfather paradox involves travelling back in time and trying to murder your grandfather before your father was conceived, thereby making it the case that you both do and do not exist.
(5) The grandfather paradox can be resolved by maintaining that there are certain things that time travellers will not do.
(6) The second time around fallacy is the fallacy of thinking that the past was first one way, and then another.
(7) Changing the past is different to affecting the past. Changing the past involves making it the case that a time is one way, and then later on is a different way. Affecting the past involves causing past events to occur.
(8) Hypertime models of time travel involve appealing to a second temporal dimension. Such models fail to show that it is possible to change the past, at least on one way of understanding what it takes to change the past.
(9) The no destination argument seeks to show that presentism is at odds with time travel based on the fact that if presentism is true there are no past times to travel to.

(10) Some philosophers maintain that time travel is improbable as it implies long strings of coincidences.

(11) Time travel seems to pose a problem for free will, but this problem can be resolved by more carefully attending to the counterfactual structure of a given time travel scenario.

(12) Closed causal loops seem to allow for events or objects to obtain without explanation.

8.8. Exercises

i. Write an inconsistent time travel story. Give it to a friend and ask them to make it consistent.

ii. Ask a friend to write an inconsistent time travel story. Make the story consistent.

iii. Draw a diagram of a consistent two-dimensional time travel story.

iv. Consider whether the moving spotlight theory from Chapter 1 is compatible with time travel to the past.

v. Defend the view that a two-dimensional model of time provides a way to change the past.

vi. Describe the difference between changing the past and affecting the past using a specific example for each.

vii. Provide two reasons in favour of the view that time travellers lack free will.

viii. Defend the idea that time travellers can reasonably deliberate about whether to kill their own grandfathers.

8.9. Glossary of Terms

Backwards Time Travel
Travel into the past.

Counterfactual
A claim about what would or might have been the case had something that actually occurred not occurred, or had something occurred that does not actually occur.

External Time
The time of a clock in an inertial frame of reference.

Forwards Time Travel
Travel into the future.

Logical Possibility
X is logically possible iff X is compatible with the laws of logic, such as the law that there are no contradictions.

Metaphysical Possibility
X is metaphysically possible iff X obtains in one of the worlds that shares the same metaphysical truths as does the actual world.

Nomological Possibility
X is nomologically possible iff X obtains in one of the worlds that has the same laws of nature as does the actual world.

Personal Time
The experienced time of the time traveller.

8.10. Further Readings

F. Arntzenius (2006) 'Time Travel: Double Your Fun', *Philosophy Compass* 1 (6): 599–616. This is a nice introduction to some of the issues raised by time travel, and extends the discussion in this chapter.

S. Bernstein (2017) 'Time Travel and the Movable Present', in John Keller, ed., *Being, Freedom, and Method: Themes from the Philosophy of Peter van Inwagen* (Oxford University Press), pp. 80–94. This is not an introductory work, but it offers a fairly accessible account of 'moving the present' models of time travel.

P. Dowe (2000) 'The Case for Time Travel', *Philosophy* 75 (3): 441–51. This is an introductory work which is best paired with William Grey's 'Troubles with Time Travel', as it is effectively a response to that paper. The two papers together provide a very nice introduction to arguments for and against the possibility of time travel.

W. Grey (1999) 'Troubles with Time Travel', *Philosophy* 74 (1): 55–70. This is an introductory work that provides a very nice overview of many of the puzzles for time travel that we introduce in this chapter.

P. Horwich (1975) 'On Some Alleged Paradoxes of Time Travel', *Journal of Philosophy* 72: 432–44. Although not an introductory work, this fairly early articulation of some of the paradoxes of time travel is accessible, and it is the first place to articulate the probabilistic problem for time travel.

D. Lewis (1976) 'The Paradoxes of Time Travel', *American Philosophical Quarterly* 13: 145–52. This is not an introductory work, but it is one of the earlier papers discussing the so-called paradoxes of time travel, and because of that it is relatively accessible. In addition, it is probably the most famous statement of, and response to, such paradoxes.

Conclusion

This book has covered a great many topics within the philosophy of time. We have looked at the nature of temporal ontology and have discussed the implications of current physics for the metaphysics of time. We have looked at the nature of temporal experience and its importance for philosophy. We have thought about the direction of time, and have considered a wide range of weird and wonderful time travel scenarios.

By way of concluding, we would like to very briefly do two things. First, we would like to identify three topics in the philosophy of time that we have not had a chance to cover, but that may intrigue the interested reader:

(i) *The difference between time and space*. Different approaches to time will take different views on just how similar or different time and space ultimately are. But every theory of time maintains that there is some difference or other between the two features of our universe. It is an interesting question as to what the difference between time and space might be.

(ii) *Temporal consciousness*. How do our conscious experiences manage to get unified over time? And how do our brains manage to unify all of the different temporal dimensions of our various sense modalities into a single experience of the present? For some, answers to these questions have metaphysical import. Work out what the best theory of temporal consciousness is and that may give you insight into the metaphysics of time. We are somewhat sceptical of the idea that temporal consciousness is quite so metaphysically laden. Nonetheless, we wish to highlight the fact that the unity of consciousness across time is an interesting area that we have not had the space to delve into here.

(iii) *Ethics and time*. One of the striking things about ethical decision-making across time is that we are not all that good at it. We are inclined to discount the future pains of others, in a manner that may not properly respect the ethical status of future individuals. The way in which time and ethics interact is becoming increasingly important as we stare down an impending climate apocalypse, one which is partly driven by our failure to reason appropriately about the future.

The second thing we would like to do is make a suggestion about where future work in the philosophy of time may be profitably directed. As we see it there are at least three interesting avenues to pursue:

(i) *The physics of time*. We can expect there to be renewed opportunities for physics and philosophy to come into close discussion as the fledgling field of quantum gravity develops. It will be interesting indeed to see what implications the ongoing development of a comprehensive account of gravity has for our understanding of time.

(ii) *Timelessness*. There is a great deal of work to be done on understanding the idea that time does not exist. If physics really does go this way, then we must begin a massive reconstruction project whereby all of the temporal notions that pervade our everyday lives are reinterpreted within a timeless setting. Agency and morality must somehow be recovered and shown to be sensible.

(iii) *The psychology of time*. Empirical psychology is a meeting point for the physics and philosophy of time. It is in the psychology of time that philosophers sometimes stand their ground, and it is the psychology of time that our best physics must ultimately be reconciled with. Our understanding of temporal experience is, however, in its infancy. As we begin to more fully understand the nature of temporal experience, we may find some answers to age-old philosophical questions about the nature of time and our place in it.

Index